Lecture Notes in Mathematics

C.I.M.E. Foundation Subseries

Volume 2308

Fondazione C.I.M.E., Firenze

C.I.M.E. stands for *Centro Internazionale Matematico Estivo*, that is, International Mathematical Summer Centre. Conceived in the early fifties, it was born in 1954 in Florence, Italy, and welcomed by the world mathematical community: it continues successfully, year for year, to this day.

Many mathematicians from all over the world have been involved in a way or another in C.I.M.E.'s activities over the years. The main purpose and mode of functioning of the Centre may be summarised as follows: every year, during the summer, sessions on different themes from pure and applied mathematics are offered by application to mathematicians from all countries. A Session is generally based on three or four main courses given by specialists of international renown, plus a certain number of seminars, and is held in an attractive rural location in Italy.

The aim of a C.I.M.E. session is to bring to the attention of younger researchers the origins, development, and perspectives of some very active branch of mathematical research. The topics of the courses are generally of international resonance. The full immersion atmosphere of the courses and the daily exchange among participants are thus an initiation to international collaboration in mathematical research.

C.I.M.E. Director (2002 – 2014)
Pietro Zecca
Dipartimento di Energetica "S. Stecco"
Università di Firenze
Via S. Marta, 3
50139 Florence
Italy
e-mail: zecca@unifi.it

C.I.M.E. Director (2015 –)
Elvira Mascolo
Dipartimento di Matematica "U. Dini"
Università di Firenze
viale G.B. Morgagni 67/A
50134 Florence
Italy
e-mail: mascolo@math.unifi.it

C.I.M.E. Secretary
Paolo Salani
Dipartimento di Matematica "U. Dini"
Università di Firenze
viale G.B. Morgagni 67/A
50134 Florence
Italy
e-mail: salani@math.unifi.it

CIME activity is carried out with the collaboration and financial support of INdAM (Istituto Nazionale di Alta Matematica)

For more information see CIME's homepage: **http://www.cime.unifi.it**

Massimo Chiappini • Vincenzo Vespri
Editors

Applied Mathematical Problems in Geophysics

Cetraro, Italy 2019

 Springer

Editors

Massimo Chiappini (iD)
Geomagnetism, Aeronomy and Environmental
Geophysics
Istituto Nazionale di Geofisica e Vulcanologia
Roma, Italy

Vincenzo Vespri (iD)
Department of Mathematics and Computer
Science "Ulisse Dini"
University of Florence
Firenze, Italy

This work was supported by INdAM (Istituto Nazionale di Alta Matematica), Istituto Nazionale di Geofisica e Vulcanologia (http://dx.doi.org/10.13039/501100008034), Strategic Initiatives for the Environment and Security (S.I.E.S.)

ISSN 0075-8434   ISSN 1617-9692 (electronic)
Lecture Notes in Mathematics
C.I.M.E. Foundation Subseries
ISBN 978-3-031-05320-7   ISBN 978-3-031-05321-4 (eBook)
https://doi.org/10.1007/978-3-031-05321-4

Mathematics Subject Classification: 86A, 65Z

This Springer imprint is published by the registered company Springer Nature Switzerland AG
The registered company address is: Gewerbestrasse 11, 6330 Cham, Switzerland

Contents

Chapter 1
Introduction to the CIME Series Volume

Applied Mathematical Problems in Geophysics

Massimo Chiappini and Vincenzo Vespri

Abstract The present volume takes its title from the CIME-EMS Summer School *"Applied Mathematical Problems in Geophysics"* held in July 2019 at Cetraro (CS) in Southern Italy. This meeting was convened 2 years after the founding of the SIES (Strategic Initiatives for the Environment and Security) initiative, a project funded by the Italian Ministry of University and Research (MIUR) and carried out jointly by the Istituto Nazionale di Geofisica e Vulcanologia (INGV) and Istituto Nazionale di Alta Matematica (INDAM). The Summer School gathered all scientists including postdoc fellows who had been involved in this research.

Featuring selected contributions based on lectures given at the CIME-EMS Summer School *"Applied Mathematical Problems in Geophysics"*, this book will provide a sound foundation for readers who intend to approach similar geophysical problems in a multidisciplinary manner.

The contributions focus on five major topics identified by the SIES as having significant societal impact: optimal control in waste management, in particular the degradation of organic waste by an aerobic biomass, by means of a mathematical model; recent developments in the mathematical analysis of subwave resonators; conservation laws in continuum mechanics, including an elaboration on the notion of weak solutions and issues related to entropy criteria; the applications of variational methods to 1-dimensional boundary value problems, in particular to light ray-tracing in ionospheric physics; and the mathematical modelling of potential electromagnetic co-seismic events associated to large earthquakes.

M. Chiappini (✉)
Geomagnetism, Aeronomy and Environmental Geophysics, Istituto Nazionale di Geofisica e Vulcanologia, Roma, Italy
e-mail: massimo.chiappini@ingv.it

V. Vespri
Università di Firenze, Florence, Italy
e-mail: vincenzo.vespri@unifi.it

The challenges continuously faced by the scientific community are quite intriguing, both in the basic and applied sciences. The paradigm behind the SIES is to adopt a synergistic and multidisciplinary approach, yielding an efficient way to identify solutions and put in place the proper actions needed to deal with the problems. It aims to resolve a series of issues related to environmental hazards and security, with the objective of realizing simulation tools based upon innovative mathematical methods and models. This initiative is inspired by the excellent results and achievements obtained by the already existing infrastructures funded by other means (PON, POR, FESR), which deal with still open problems that are not easily solved solely with the competences present at INGV. The close cooperation with INDAM is the key to bridging the gaps represented by the deep knowledge of mathematical tools needed to obtain the solutions of complex problems.

The SIES adopts a modular structure to target environmental, industrial and civil issues. The five modules are as follows:

1. The study of the evolution of a bio-reactor through models of production of leachate during waste decomposition.
2. The prediction of the direction of waves due to sea conditions, to be validated with radar techniques.
3. The multidimensional characterization of elastic wave propagation in the ground when using metamaterials.
4. The characterization of the electromagnetic sources inside the Earth based upon techniques of tensor gradiometry and the properties of characteristic tensors.
5. The analytical formulation of ray-tracing of electromagnetic waves in an ionized environment aimed at the protection of national borders.

Module No. 1 has applications with a significant industrial and environmental impact, including the generation of secure, clean and efficient energy. The outcome of processing organic waste to produce energy is enormous, with great potential for funding opportunities in applied science programs issued by national and European agencies.

Module No. 2 offers the possibility of applications in the field of harbor security (La Spezia Harbor is currently adopting a technological solution designed by INGV), as well as opportunities related to renewable energy sources (tidal energy).

Module No. 3 represents the topic with the highest potential. New trends in the study of metamaterials aimed at isolating infrastructures from vibrations are affecting both industrial and social sectors. Enhancing the expertise on cloaking in Italy improves the know-how of a sector with a growing strategic importance. Applications in this field would attract various dual users, from civilian to military. Potential stakeholders are, inter alia, the Department of Civil Protection of the Office of the Prime Minister, and the Ministry of Defense.

While modules 1 to 3 are based upon applied research, modules No. 4 and 5 are more oriented toward basic research. In module *No. 4*, studying electromagnetic signals as events associated to the preparatory phases of a large earthquake is considered a national duty for a developed country. The identification and characterization of these signals can save lives in the future, allowing the development of appropriate early warning systems.

Module No. 5 deals with a detailed 3D characterization of the Earth's ionosphere. A deep knowledge of the structural properties of Maxwell, Navier–Stokes, and Fermat equations is required. The capability to analytically formulate the ray-tracing of electromagnetic waves in an ionized environment has a number of advantages, with applications both in the military framework and in the field of homeland security.

The SIES effort has an additional academic impact: it promotes industrial mathematics as a sector not yet fully developed in Italy, with a high potential for employment of young mathematicians and geophysicists.

The content of this book reflects the above modules. Readers will be able to get a broader perspective of the various topics by consulting the bibliographies provided at the end of each lecture.

The editors of the present volume wish to acknowledge the Istituto Nazionale di Alta Matematica (INDAM) and the Istituto Nazionale di Geofisica e Vulcanologia (INGV) for their valuable support provided during the conduct of the research. The CIME Foundation provided the unique opportunity to hold the Summer School in an excellent location where senior and young scientists had the chance to fruitfully interact, and also funded the work that led to the publication of the present book.

Chapter 2
Optimal Control Strategies for Composting Processes in Biocells with L_1−Type and L_2−Type Cost Objectives

Giorgio Martalò, Cesidio Bianchi, Bruno Buonomo, Massimo Chiappini, and Vincenzo Vespri

Abstract We present a finite horizon optimal control problem for composting in biocell. The problem is based on a mathematical model, which describes the degradation of organic waste by an aerobic biomass. Solubilization of insoluble substrate and biomass decay phenomena are also taken into account. The degradation is controlled by monitoring the effects of oxygen concentration in the cell. The optimal strategy is determined by minimizing the total costs given by the final size of soluble substrate and the operating cost. We consider two different cases, where the operating cost is modeled by linear and quadratic function of the effort, respectively. It is shown that the total costs over the considered time interval are higher in the linear case than in the quadratic case. Moreover, when compared to the quadratic case, the linear case results in a higher reduction of soluble substrate concentration but it requires also a higher effort to control the degradation.

G. Martalò (✉)
Dipartimento di Scienze Matematiche Fisiche e Informatiche, Università di Parma, Parma, Italy
e-mail: giorgio.martalo@unipr.it

C. Bianchi · M. Chiappini
Geomagnetism, Aeronomy and Environmental Geophysics, Istituto Nazionale di Geofisica e Vulcanologia, Roma, Italy
e-mail: cesidio.bianchi@ingv.it; massimo.chiappini@ingv.it

B. Buonomo
Dipartimento di Matematica e Applicazioni Renato Caccioppoli, Università di Napoli Federico II, Napoli, Italy
e-mail: buonomo@unina.it

V. Vespri
Dipartimento di Matematica e Informatica Ulisse Dini, Università di Firenze, Firenze, Italy
e-mail: vincenzo.vespri@unifi.it

© The Author(s), under exclusive license to Springer Nature Switzerland AG 2022
M. Chiappini, V. Vespri (eds.), *Applied Mathematical Problems in Geophysics*,
C.I.M.E. Foundation Subseries 2308, https://doi.org/10.1007/978-3-031-05321-4_2

5

2.1 Introduction

Waste management is a current issue of great interest, especially for the identification of suitable intervention policies that allow to overcome some criticisms, like the increasing requirement of new stocking sites, the risk of soil and aquifers contamination, air pollution, formation and diffusion of leachate [8, 13, 21].

For such reason, nowadays a landfill is conceived as a biological active environment, called bioreactor [20], where the control of the biodegradation phenomenon allows to improve the plant performance, in terms of reduced waste amount and of compost and biogas production [9, 13].

In this framework, mathematical models for anaerobic [14, 27] and aerobic [12, 16, 25, 26] digestion processes and some related control problems [4, 6, 17, 18, 24] provide an useful tool to suggest suitable intervention policies to be adopted and prevent the use of poorly efficient strategies. Moreover, the purely theoretical approach allows to face and discuss many different scenarios, choosing from time to time new processes to be controlled and goals to be achieved.

As regards the anaerobic digestion, in [4] the biogas production in a continuous filled bioreactor is maximized by controlling the feeding process, while in [6] and [24] the optimal leachate recirculation strategy is determined to control the operating costs in a given time range and to reduce the total amount of substrate in minimal time, respectively.

As concerns the aerobic process, some optimal control problems for composting in biocells have been recently proposed [17, 18]. The main purpose of such control problems is to provide some qualitative indications about aeration strategies that allow to maintain appropriate levels of oxygen in the cell (and hence the aerobic feature of the process), in addition to the maximization of compost production. Moreover, some realistic features, like the inhibition effects due to over-aeration and the economic cost of the aeration operating, have been taken into account and their influence on the optimal strategy has been discussed, also in a real world scenario [18].

A different model has been proposed in [19] to describe the composting process in a biocell. Moreover, in [19] a time optimal control problem for the reduction of solid waste in minimal time has been discussed. Such model preserves some distinctive features, as already pointed out in [17, 18], like a two component (soluble and insoluble) structure of the substrate and the main transformation phenomena: (i) degradation of the soluble substrate by an aerobic bacterial population; (ii) hydrolysis, i.e. solubilization of the insoluble component; (iii) biomass decay, that produces new insoluble substrate.

We stress that in these modeling approaches the control is described through the effects of oxygen on the degradation term, where the oxygen level (and hence its effects) can be regulated by a suitable injection/aspiration system [23].

As further contribution to the analysis given in [19], here we propose an optimal control problem on a finite horizon. The main aim is to minimize in a fixed time range an optimality criterion, which averages the reduction of the soluble component at the final time and the economic cost of the control operating. The first term can be used as a maturation index in composting, as suggested in [5, 7]. The second term is described first by a linear function of the effort and then by a quadratic function. As it has been pointed out many times [3, 6, 15], there is not a shared standard form of objective to be preferred, also because of the different cost sources to be modeled [1, 10]. The main advantage of a quadratic function lies in its mathematical simplicity and in the convexity of the corresponding Hamiltonian, which guarantees the uniqueness of the minimum. Alternatively, the linear term is easier to justify from a modeling point of view but it may lead to a more complicated analysis, often due to the occurrence of bang-bang controls.

The paper is organized as follows: after recalling the mathematical model, we state the optimal control problem and reformulate it by means of Pontryagin's theory [22] in Sect. 2.2; in Sect. 2.3 the optimal time profiles are obtained numerically and the results obtained by using strategies with linear and quadratic cost, respectively, are compared. Concluding remarks are given in Sect. 2.4.

2.2 The Optimal Control Problem

2.2.1 Formulation of the Problem

We set an optimal control problem on a fixed time range $[0, T]$ to improve the performance of a composting biocell, i.e. a closed system for waste degradation where no ingoing or outgoing flow of material is taken into account. The main aim of the control problem is to minimize the total costs due to soluble substrate concentration at final time T and the time averaged operating cost. This latter is due to aeration and aspiration, which allow to regulate the level of oxygen in the biocell atmosphere. The effect of the oxygen on the degradation of soluble substrate (say, s) is described by the control variable u. The cost is modeled as a linear or quadratic function of the control variable u. We consider also the hydrolytic process of solubilization of the insoluble component of the substrate (say, i) and the decay phenomenon of the bacterial biomass (say, x), that degrades the soluble substrate.

Our goal is to determine the optimal control $u(t)$, $t \in [0, T]$, in the admissible control set

$$\mathcal{U} = \{v : [0, T] \longrightarrow [u_{\min}, u_{\max}], \, v \text{ Lesbesgue measurable}\}, \quad (2.1)$$

minimizing the objective functional

$$J_k[u] = \left(\frac{s(T)}{m}\right)^2 + \frac{\alpha}{T} \int_0^T u^k(\tau)d\tau, \quad (2.2)$$

where $k = 1, 2$, the quantity $m > 0$ is the constant total mass in the cell and the parameter $\alpha > 0$ averages the two contributions in the optimality criterion.

We consider a quadratic term for soluble substrate concentration at final time T in order to emphasize/demphasize large/small deviations with respect to the total consumption, that represents the most desirable scenario.

Linear and quadratic cases are reproduced when $k = 1$ and $k = 2$, respectively (they will be indicated sometimes as L_1 and L_2, respectively). The optimal strategy to be adopted depends mainly on the minimization of the final soluble substrate when α is small, while it is mainly determined by the minimization of the operating cost for large values of α.

The time evolution of the variables s, i and x is governed, under perfect mixing conditions, by the following system of ordinary differential equations [19]

$$\frac{ds}{dt} = -\mu \frac{s}{s + c_s} ux + c_h i \tag{2.3}$$

$$\frac{di}{dt} = -c_h i + bx \tag{2.4}$$

$$\frac{dx}{dt} = -bx + \mu \frac{s}{s + c_s} ux , \tag{2.5}$$

where t is the time variable, $\mu > 0$ is the maximal growth rate, $c_s > 0$ is the half-saturation constant, $c_h > 0$ is the hydrolytic coefficient and $b > 0$ is the biomass decay constant.

The control variable u is supposed to be non-negative and takes values in a given range $[u_{min}, u_{max}]$ $(0 \leq u_{min} < u_{max})$. The optimal operational concentration of the oxygen in the cell atmosphere, ω, is around 10% and a range around this value, say $5\% = \omega_{min} \leq \omega \leq \omega_{max} = 15\%$, guarantees the aerobic feature of the degradation [9]. We model the effects of oxygen on the degradation term by a Monod response function of the oxygen variable ω as follows

$$f(\omega) = \frac{\omega}{\omega + c_\omega} , \tag{2.6}$$

where the half saturation constant c_ω is set equal to 2% [26].

From relation (2.6), it follows that the operational range $[\omega_{min}, \omega_{max}]$ corresponds to the set of admissible values of the control

$$[u_{min}, u_{max}] = [5/7, 15/17] \simeq [0.714, 0.882] . \tag{2.7}$$

It can be easily checked that the conservation of total mass holds

$$M(t) = s(t) + i(t) + x(t) = s(0) + i(0) + x(0) = M(0) =: m , \tag{2.8}$$

and therefore the system (2.5) reduces to

$$\frac{ds}{dt} = -\mu \frac{s}{s+c_s} u(m-s-i) + c_h i \tag{2.9}$$

$$\frac{di}{dt} = -c_h i + b(m-s-i), \tag{2.10}$$

where the variable $x(t)$ is replaced by $m - s(t) - i(t)$.

The minimization problem can be investigated in the positively invariant set

$$\mathcal{F} = \{(s, i) \in [0, m] \times [0, m] \text{ such that } s + i \le m\}, \tag{2.11}$$

since any solution of (2.9)-(2.10) with initial condition in $\mathcal{F}$ remains in $\mathcal{F}$ for any positive time.

2.2.2 Analysis of the OC Problem

We reformulate the problem by means of Pontryagin's minimum principle [22].

We introduce the Hamiltonian function

$$\mathcal{H}_k(s, i, \lambda_s, \lambda_i, u, t) = \frac{\alpha}{T} u^k(t) + (m - s - i) \left(b\lambda_i - \mu \frac{s}{s+c_s} u\lambda_s \right) c_h i (\lambda_s - \lambda_i), \tag{2.12}$$

where $k = 1, 2$, and the time dependent *adjoint variables* (λ_s, λ_i) solve the *adjoint system* of ordinary differential equations

$$\dot{\lambda}_s = -\frac{\partial \mathcal{H}_k}{\partial s} = \frac{\mu}{(s+c_s)^2} [c_s(m-s-i) - s(s+c_s)] u\lambda_s + b\lambda_i \tag{2.13}$$

$$\dot{\lambda}_i = -\frac{\partial \mathcal{H}_k}{\partial i} = -\left(\mu \frac{s}{s+c_s} u + c_h \right) \lambda_s + (b + c_h)\lambda_i, \tag{2.14}$$

with given *transversality condition* at final time T

$$(\lambda_s(T), \lambda_i(T)) = (2s(T)/m^2, 0). \tag{2.15}$$

We introduce the function

$$\phi_k = \frac{\partial \mathcal{H}_k}{\partial u} = \frac{\alpha}{T} u^{k-1} - \mu \frac{s}{s+c_s} (m-s-i)\lambda_s, \quad k = 1, 2, \tag{2.16}$$

that provides the following characterization of the optimal control

$$
u \begin{cases} = u_{\min} \\ \in (u_{\min}, u_{\max}) \\ = u_{\max} \end{cases} \quad \text{if} \quad \phi_k \begin{cases} > 0 \\ = 0 \\ < 0 \end{cases} . \tag{2.17}
$$

As it is well known [11], if $\phi_k(t) = 0$ for any $t \in [t_1, t_2] \subset [0, T]$ $(t_1 < t_2)$, then the control is called *singular* and assumes values between $u_{\min}$ and $u_{\max}$. If the optimal control is not singular, then it must be constant or piecewise constant and is called *bang* or *bang-bang*, respectively.

We observe also that a possible singular optimal control can be formulated in terms of the state and adjoint variables when the cost term is a quadratic function, since the condition $\phi_2 = 0$ gives

$$
u = \mu \frac{T}{2\alpha} \frac{s}{s + c_s} (m - s - i)\lambda_s(t) . \tag{2.18}
$$

We discuss now the presence of a minimizing singular control when the cost term is linear $(k = 1)$.

Let $u(t)$ be a singular control for $t \in [t_1, t_2] \subset [0, T]$ and $(s(t), i(t))$ be the corresponding solution in such interval; the problem order is the smallest number q such that the $2q$-th derivative

$$
\frac{d^{2q}}{dt^{2q}} \frac{\partial \mathcal{H}_1}{\partial u}(s, i, \lambda_s, \lambda_i, u, t) = \frac{d^{2q}}{dt^{2q}} \phi_1(s, i, \lambda_s, \lambda_i, u, t) \tag{2.19}
$$

explicitly contains the control variable u (if no derivative satisfies this condition then $q = \infty$).

In our analysis, from (2.16) with $k = 1$, it follows that

$$
\phi_1 = \frac{\alpha}{T} - \mu \frac{s}{s + c_s} (m - s - i)\lambda_s , \tag{2.20}
$$

and its second order derivative

$$
\ddot{\phi}_1 = \left\{ \lambda_s b \frac{s^2}{(s + c_s)^2} + \lambda_s c_h i \left[-2 \frac{s c_s}{(s + c_s)^4} - \frac{c_s^2}{(s + c_s)^4} \right] (m - s - i) \right. \tag{2.21}
$$

$$
\left. \lambda_i b \frac{s}{s + c_s} \left[\frac{c_s}{(s + c_s)^2} (m - s - i) - \frac{s}{s + c_s} \right] \right\} \mu^2 (m - s - i)u \tag{2.22}
$$

$$
+ \mathcal{R}(s, i, \lambda_s, \lambda_i) \tag{2.23}
$$

explicitly contains u. The quantity $\mathcal{R}$ is a collection of terms, which do not contain u explicitly. It follows that our problem is of order 1.

A sufficient condition for a singular control to be minimizer is provided by the *Legendre-Clebsch condition* [11], that reads as

$$\mathcal{L}(s, i, \lambda_s, \lambda_i) = \frac{\partial}{\partial u} \frac{d^2}{dt^2} \frac{\partial \mathcal{H}_1}{\partial u} = \frac{\partial}{\partial u} \ddot{\phi}_1 \leq 0 \tag{2.24}$$

for problems of order 1.

In our case such derivative is given by

$$\mathcal{L}(s, i, \lambda_s, \lambda_i) \tag{2.25}$$

$$= \left\{ \lambda_s b \frac{s^2}{(s + c_s)^2} + \lambda_s c_h i \left[-2 \frac{s c_s}{(s + c_s)^4} - \frac{c_s^2}{(s + c_s)^4} \right] (m - s - i) \right. \tag{2.26}$$

$$\left. \lambda_i b \frac{s}{s + c_s} \left[\frac{c_s}{(s + c_s)^2} (m - s - i) - \frac{s}{s + c_s} \right] \right\} \mu^2 (m - s - i) \tag{2.27}$$

In presence of a singular control, the function given in (2.20) is such that

$$\phi_1 = \dot{\phi}_1 = 0, \tag{2.28}$$

where

$$\dot{\phi}_1 = \mu(m - s - i) \left[b \frac{s}{s + c_s} (\lambda_s - \lambda_i) - \frac{c_s}{(s + c_s)^2} c_h i \lambda_s \right] ; \tag{2.29}$$

therefore, for any configuration in the interior of $\mathcal{F}$, the derivative in (2.27) can be rewritten as

$$\mathcal{L}(s, i, \lambda_s, \lambda_i) = 2 \frac{\alpha}{T} \frac{c_s}{(s + c_s)^2} [bs(m - s - i) - c_h i (2m - 3s - 2i)] . \tag{2.30}$$

We observe that the Legendre-Clebsch condition (2.24) is fulfilled in the subset

$$\mathcal{F}_- = \{(s, i) \in \mathcal{F} \text{ such that } bs(m - s - i) - c_h i (2m - 3s - 2i) < 0\} \subset \mathcal{F}, \tag{2.31}$$

and hence any singular control involving states in $\mathcal{F}_-$ is guaranteed to be a minimizer.

2.3 Numerical Results

In this section we solve the minimization problem for linear and quadratic cost functionals, by using a numerical technique based on a gradient descent method [2].

We consider the thermophilic phase of composting [9], lasting about 3 days (72 h); in this phase, the temperature is supposed to be maintained at around 55 °C.

We consider the following parameters given in [16]:

$$\mu = 0.36\,h^{-1}, \; c_s = 0.0772\,Kg, \tag{2.32}$$

$$c_h = 1.764 \times 10^{-3}h^{-1}, \; b = 0.1915\,h^{-1}. \tag{2.33}$$

Such values correspond to a cell temperature of 55 °C.

For illustrative purposes we consider the following initial conditions

$$s(0) = 2200\,Kg, \; i(0) = 3600\,Kg, \; x(0) = 2Kg. \tag{2.34}$$

2.3.1 Comparison of Strategies for Linear and Quadratic Cost Terms

We first compare the optimal strategies for linear and quadratic cost terms, when the two contributions in the objective functional have the same weight in the adopted strategy ($\alpha = 1$).

We observe (see Fig. 2.1d) that the optimal control in the linear case is given by

$$u_1 = \begin{cases} u_{\max} & \text{when } 0 \leq t \leq t_0 \\ \text{singular} & \text{when } t_0 < t < t_1 \\ u_{\min} & \text{when } t_1 \leq t \leq T, \end{cases} \tag{2.35}$$

where $t_0 \simeq 16.18$ and $t_1 \simeq 17.23$, while in the quadratic case the optimal control is

$$u_2 = \begin{cases} \text{singular} & \text{when } 0 < t < t_2 \\ u_{\min} & \text{when } t_2 \leq t \leq T, \end{cases} \tag{2.36}$$

where $t_2 \simeq 44.57$.

In both cases the control is singular on a suitable subset of $[0, T]$, even if the control trend in the linear case seems very close to a bang-bang profile. Moreover, in any case, the optimal control in the very last time range is given by the constant control $u_{\min}$ and the main difference between the two cases is that the first time instant at which the control takes its minimal value. This means that starting from

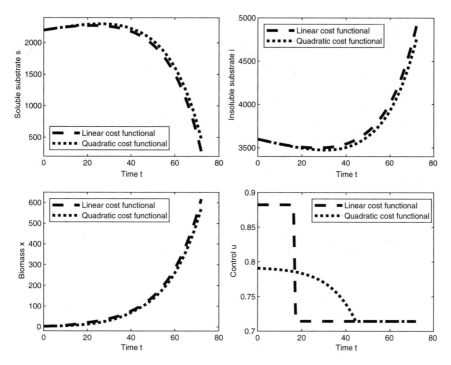

Fig. 2.1 State variables profiles (first three panels) and optimal control (last panel) for the test case (2.32)–(2.34) and $\alpha = 1$, when the cost is modeled by a linear or quadratic function

a given time it is optimal to consider the minimal control guaranteeing the aerobic feature of the process and no additional effort is required.

As concerns the state variables, we notice that they present a similar behavior in both cases (see Fig. 2.1a–c). In particular, soluble substrate slightly increases at the beginning, since the action of a small concentration of bacteria is not sufficient to balance the effects of hydrolysis. As soon as the bacterial concentration increases, the soluble substrate concentration reduces and reaches a very small value at the final time T. The opposite trend governs the insoluble component, while the biomass concentration strictly increases at any time.

The decrease of soluble concentration is compatible with the biological assumptions stated above, since a significant reduction of such component of substrate is expected in the thermophilic phase. The increase of the other variables follows from the conservation of total mass (2.8).

Although the time profiles of the variables are qualitatively comparable for linear and quadratic cases, the corresponding state configurations at final time T are significantly different from a quantitative point of view. In fact, as shown in Table 2.1, the optimal control in the linear case leads to a higher reduction of soluble

Table 2.1 Percentages of soluble substrate reduction, insoluble component increase and additional required effort for L_1 and L_2 cost terms in the test case (2.32)–(2.34) and $\alpha = 1$

	$\dfrac{s(T) - s(0)}{s(0)}$	$\dfrac{i(T) - i(0)}{i(0)}$	$\dfrac{I_1 - I_1^{\min}}{I_1^{\min}}$
Linear cost term	-87.3%	$+36.3\%$	$+5.46\%$
Quadratic cost term	-78.3%	$+32.0\%$	$+5.04\%$

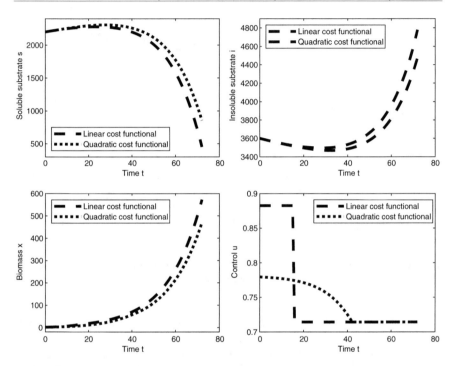

Fig. 2.2 State variables profiles (first three panels) and optimal control (last panel) for the test case (2.32)–(2.34) and $\alpha = 1.5$ and linear and quadratic cost terms

substrate than the one obtained in the quadratic case. Moreover, in the quadratic case the mean effort, given by

$$I_1 = \frac{1}{T} \int_0^T u(\tau) d\tau \,, \tag{2.37}$$

takes a lower value than in the linear case. As a consequence, the same happens to the difference between the mean effort and the minimal effort $I_1^{\min} = I_1(u = u_{\min})$.

The qualitative and quantitative features discussed above are emphasized for increasing values of α, as shown in Fig. 2.2 and Table 2.2 where $\alpha = 1.5$.

The optimal time profiles of the control in both the L_1 and L_2 cases are the same as (2.35) and (2.36), but here $t_0 = 14.95$, $t_1 = 15.93$ and $t_2 = 42.19$. As expected,

Table 2.2 Percentages of soluble substrate reduction, insoluble component increase and additional required effort for L_1 and L_2 cost terms in the test case (2.32)–(2.34) and $\alpha = 1.5$.

	$\dfrac{s(T) - s(0)}{s(0)}$	$\dfrac{i(T) - i(0)}{i(0)}$	$\dfrac{I_1 - I_1^{\min}}{I_1^{\min}}$
Linear cost term	−79.5%	+32.8%	+5.04%
Quadratic cost term	−61.1%	+24.4%	+3.98%

the percentage of soluble substrate reduction and additional effort are lower than the case $\alpha = 1$, since now the minimization of the operating cost term has a dominant role in determining the decision policy to be adopted. We observe also a higher quantitative discrepancy between linear and quadratic cases than the one observed in the test case with $\alpha = 1$.

2.3.2 Analysis for Varying Balance Between Costs

As shown in the previous subsection, some differences between optimal profiles in L_1 and L_2 cases are emphasized for different choices of the parameter α.

We remind here that such parameter averages the role of the two contributions in the objective functional. More precisely, the optimal strategy is mainly determined by the minimization of soluble substrate at time T when α is small; the minimization of the (linear or quadratic) operating cost term is the key factor in the individuation of the optimal control when α is large.

We discuss in this subsection the main differences between L_1 and L_2 cases for varying α, by investigating the trend of some performance indices. In particular, we consider

– the value, denoted by I_0, of the corresponding objective functional $J_k[u]$, $k = 1, 2$, given in (2.2);
– the mean effort I_1 introduced in (2.37);
– the soluble substrate concentration at final time T: $I_2 = s(T)/m$;
– the first time instant at which the control assumes its minimal value constantly: $I_3 = \{\min \tau \text{ such that } \tau \in [0, T] \text{ and } u(\tau) = u_{\min}\}$.

We observe that the value I_0 of the objective functional increases monotonically in both the linear and quadratic case (see Fig. 2.3a). The total costs in presence of a linear operating cost term are higher than the ones obtained with a quadratic function.

The discrepancy between the two cases is quantified by

$$i_0 = \frac{I_0(k = 1) - I_0(k = 2)}{I_0(k = 2)}, \qquad (2.38)$$

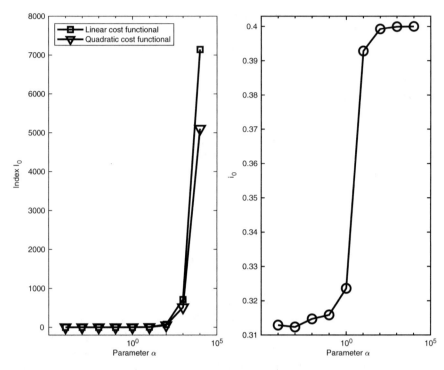

Fig. 2.3 Trend of the performance index I_0 for L_1 and L_2 cases for the test scenario (2.32)–(2.34) and parameter α varying in the range 10^{-4}–10^4 (left panel) and the percentage difference i_0 in (2.38) (right panel)

corresponding to the percentage difference between the two cases. We observe that the total costs in the linear case are 30–40% higher than those of the quadratic case (see Fig. 2.3b).

The role of each term in the objective functional in such difference is discussed by means of indices I_1 and I_2, whose trends are reported in Figs. 2.4 and 2.5, respectively. It is shown that, although the strategies corresponding to linear and quadratic cost terms are deeply different, they both lead to similar results when the parameter α is sufficiently small or it is sufficiently large. In fact, when α is small the minimization of the operating cost term has a negligible role and in both cases the strategies to be adopted have the aim to reduce the soluble substrate concentration, regardless of limiting the operating costs. On the contrary, when α is large, the minimization of the costs is the key factor in determining the optimal control and, from a certain value of α on, any control operating is considered too expensive to be taken into account. Therefore, in both cases, it is optimal to consider a constant control with minimal value $u_{\min}$.

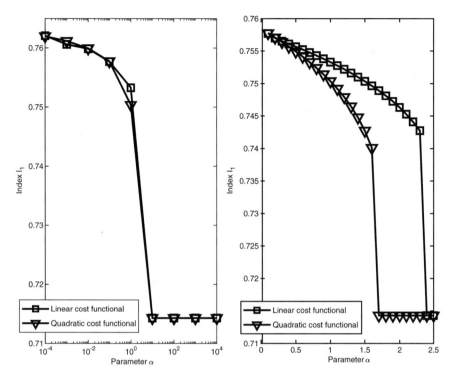

Fig. 2.4 Trend of the performance index I_1 for L_1 and L_2 cases for the test scenario (2.32)–(2.34) and parameter α varying in the range 10^{-4}–10^4 (left panel) and in the interval $0.1-2.5$ (right panel)

The range of interest for parameter α is $0.1 \leq \alpha \leq 2.5$, where the differences between linear and quadratic cases are appreciable (see panels (b) in Figs. 2.4 and 2.5).

As already observed in the previous subsection, the strategy minimizing the linear cost leads to a larger reduction of soluble substrate compared to the quadratic case (see Fig. 2.5b). Moreover, when the cost term is quadratic, the effort required to control the system and optimize the objective functional is lower when compared to the linear case (see Fig. 2.4b). The differences between L_1 and L_2 cases are more and more significant for increasing values of the parameter α.

As last remark, we notice also that the first value of α at which the control assumes the minimal value constantly is not the same for linear and quadratic cost objectives (see Fig. 2.6) and the constant control strategy, $u(t) = u_{min}$ for any $t \in [0, T]$, occurs "later" in the linear case (see Fig. 2.6b).

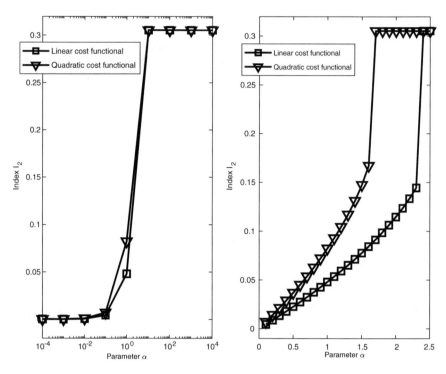

Fig. 2.5 Trend of the performance index I_2 for L_1 and L_2 cases for the test scenario (2.32)–(2.34) and parameter α varying in the range 10^{-4}–10^4 (left panel) and in the interval 0.1–2.5 (right panel)

2.4 Conclusions

We have proposed and discussed a finite horizon optimal control problem for a biological system, in which an aerobic bacterial population degrades the soluble component of the substrate. The solubilization of the insoluble component and a biomass decay phenomenon are also taken into account.

The objective functional expresses the total costs due to the minimization of the soluble substrate component at final time and to the minimization of operating costs. This latter term has been modeled with a linear and quadratic function of the control variable.

We have observed that the optimal profiles of the state variables for the linear and the quadratic cases are qualitatively similar. The two cases significantly differ from a quantitative point of view. Such discrepancy is evident in the value of the objective function, since total costs in the linear case are 30–40% higher than ones of the quadratic case. Moreover, in the linear case, we observe a major reduction of soluble substrate and a higher effort is required to control the system.

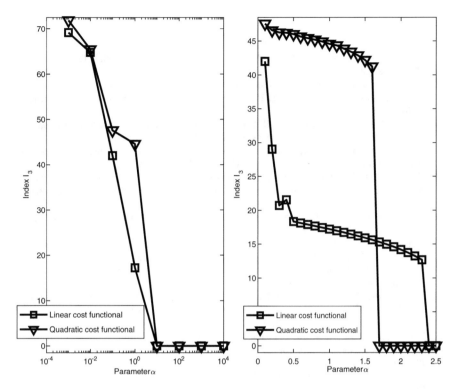

Fig. 2.6 Trend of the performance index I_3 for L_1 and L_2 cases for the test scenario (2.32)–(2.34) and parameter α varying in the range 10^{-4}–10^4 (left panel) and in the interval $0.1-2.5$ (right panel)

Such quantitative differences are more and more evident for increasing values of the parameter α, when the operating cost minimization plays a more and more influential role in determining the optimal strategy. An analysis for varying α has confirmed such remark, up to a certain value of α. From this threshold on, any operating is considered too expensive to be adopted in both cases, and the optimal control is constantly equal to the minimal admissible value, guaranteeing the aerobic feature of the degradation.

In view of the uncertainty concerning the suitable way to describe the operating cost for composting in biocell, we believe that this theoretical approach could be incorporated in future practice to provide insight into possible composting strategies.

Acknowledgments This work was performed in the framework of INdAM-INGV joint project *Strategic Initiatives for the Environment and Security* (SIES) supported by MIUR (Progetto Premiale FOE2014).

References

1. L. Amand, G. Olsson, B. Carlsson, Aeration control–a review. Water Sci. Technol. **67**(11), 2374–2398 (2013)
2. S. Anita, V. Capasso, V. Arnautu, *An Introduction to Optimal Control Problems in Life Sciences and Economics: From Mathematical Models to Numerical Simulation with MATLAB* (Springer, Berlin, 2011)
3. O. Bara, S.M. Djouadi, J.D. Day, S. Lenhart, Immune therapeutic strategies using optimal controls with L^1 and L^2 type objectives. Math. Biosci. **290**, 9–21 (2017)
4. T. Bayen, O. Cots, P. Gajardo, Analysis of an optimal control problem related to anaerobic digestion process. J. Optim. Theory Appl. **178**(2), 627–659 (2018)
5. M.P. Bernal, C. Paredes, M.A. Sanchez-Monedero, J. Cegarra, Maturity and stability parameters of composts prepared with a wide range of organic wastes. Bioresour. Technol. **63**(1), 91–99 (1998)
6. M. Bisi, M. Groppi, G. Martalò, R. Travaglini, Optimal control of leachate recirculation for anaerobic processes in landfills. Discrete Cont. Dyn. B **26**(6), 2957–2976 (2021)
7. M. Borisov, N. Dimitrova, V. Beschkov, Stability analysis of a bioreactor model for biodegradation of xenobiotics. Comput. Math. Appl. **64**(3), 361–373 (2012)
8. S. Bozkurt, L. Moreno, I. Neretnieks, Long-term processes in waste deposits. Sci. Total Environ. **250**(1–3), 101–121 (2000)
9. L. Cooperband, The art and science of composting. Center for Integrated Agricultural Systems (2002). Available at https://www.cias.wisc.edu/wp-content/uploads/2008/07/artofcompost.pdf. Accessed on February 2019
10. B. Davidson, R.W. Bradshaw, A steady state optimal design of artificial induced aeration in polluted streams by the use of Pontryagin's minimum principle. Water Resour. Res. **6**(2), 383–397 (1970)
11. D. Grass, J.P. Caulkins, G. Feichtinger, G. Tragler, D.A. Behrens, *Optimal Control of Nonlinear Processes* (Springer, Berlin, 2008)
12. H.V.M. Hamelers, A mathematical model for composting kinetics (2001). Available at http://library.wur.nl/WebQuery/wurpubs/fulltext/193815. Accessed on Mar 2018
13. R.T. Haug, *The Practical Handbook of Compost Engineering* (Lewis Publishers, Boca Raton, 1993)
14. A. Husain, Mathematical models of the kinetics of anaerobic digestion – a selected review. Biomass Bioenerg. **14**(5–6), 561–571 (1998)
15. U. Ledzewicz, T. Brown, H. Schättler, Comparison of optimal controls for a model in cancer chemotherapy with L_1- and L_2-type objectives. Optim. Methods Softw. **19**(3–4), 339–350 (2004)
16. Y.P. Lin, G.H. Huang, H.W. Lu, L. He, Modeling of substrate degradation and oxygen consumption in waste composting processes. Waste Manag. **28**(8), 1375–1385 (2008)
17. G. Martalò, C. Bianchi, B. Buonomo, M. Chiappini, V. Vespri, Mathematical modeling of oxygen control in biocell composting plants. Math. Comput. Simul. **177**, 105–119 (2020)
18. G. Martalò, C. Bianchi, B. Buonomo, M. Chiappini, V. Vespri, On the role of inhibition processes in modeling control strategies for composting plants. SEMA SIMAI Springer Series **21**, 125–145 (2020)
19. G. Martalò, C. Bianchi, B. Buonomo, M. Chiappini, V. Vespri, A minimum time control problem for aerobic degradation processes in biocell composting plants. Optim. Contr. Appl. Met. **41**(4), 1251–1266 (2020)
20. J. Pacey, D. Augenstein, R. Morck, D. Reinhart, R. Yazdani, The bioreactor landfill-an innovation in solid waste management. MSW (1999). Available at https://citeseerx.ist.psu.edu/viewdoc/download?doi=10.1.1.460.2984&rep=rep1&type=pdf. Accessed on Mar 2021
21. C. Polprasert, *Organic Waste Recycling* (John Wiley and Sons Ltd., Hoboken, 1989)
22. L.S. Pontryagin, *Mathematical Theory of Optimal Processes* (CRC Press, Boca Raton, 1987)

23. E.C. Rada, M. Ragazzi, S. Villotti, V. Torretta, Sewage sludge drying by energy recovery from OFMSW composting: preliminary feasibility evaluation. Waste Manag. **34**(5), 859–866 (2014)
24. A. Rapaport, T. Bayen, M. Sebbah, A. Donoso-Bravo, A. Torrico, Dynamical modeling and optimal control of landfills. Math. Models Methods Appl. Sci. **26**(05), 901–929 (2016)
25. B. Seng, H. Kaneko, Simulation of windrow composting for organic solid wastes, in *International Conference on Chemical, Biological and Environment Sciences (ICCEBS–2011)*, Bangkok (2011)
26. B. Seng, R.A. Kristanti, T. Hadibarata, K. Hirayama, K. Katayama-Hirayama, H. Kaneko, Mathematical model of organic substrate degradation in solid waste windrow composting. Bioprocess Biosyst. Eng. **39**(1), 81–94 (2016)
27. V.A. Vavilin, S.V. Rytov, L.Y. Lokshina, S.G. Pavlostathis, M.A. Barlaz, Distributed model of solid waste anaerobic digestion: effects of leachate recirculation and pH adjustment. Biotechnol. Bioeng. **81**(1), 66–73 (2003)

Chapter 3
Wave Interaction with Subwavelength Resonators

Habib Ammari, Bryn Davies, Erik Orvehed Hiltunen, Hyundae Lee, and Sanghyeon Yu

Abstract The aim of this review is to cover recent developments in the mathematical analysis of subwavelength resonators. The use of sophisticated mathematics in the field of metamaterials is reported, which provides a mathematical framework for focusing, trapping, and guiding of waves at subwavelength scales. Throughout this review, the power of layer potential techniques combined with asymptotic analysis for solving challenging wave propagation problems at subwavelength scales is demonstrated.

3.1 Introduction

The ability to focus, trap, and guide the propagation of waves on subwavelength scales is of fundamental importance in physics. Systems of subwavelength resonators have, in particular, been shown to have desirable and sometimes remarkable properties thanks to their tendency to interact very strongly with waves on small length scales [26, 31, 33]. A subwavelength resonator is a cavity with material

H. Ammari (✉)
Department of Mathematics, ETH Zürich, Zürich, Switzerland
e-mail: habib.ammari@math.ethz.ch

B. Davies
Department of Mathematics, Imperial College London, London, UK
e-mail: bryn.davies@imperial.ac.uk

E. O. Hiltunen
Department of Mathematics, Yale University, New Haven, CT, USA
e-mail: erik.hiltunen@yale.edu

H. Lee
Department of Mathematics, Inha University, Incheon, South Korea
e-mail: hdlee@inha.ac.kr

S. Yu
Department of Mathematics, Korea University, Seoul, South Korea
e-mail: sanghyeon_yu@korea.ac.kr

© The Author(s), under exclusive license to Springer Nature Switzerland AG 2022
M. Chiappini, V. Vespri (eds.), *Applied Mathematical Problems in Geophysics*,
C.I.M.E. Foundation Subseries 2308, https://doi.org/10.1007/978-3-031-05321-4_3

parameters that are greatly different from the background medium and which experiences resonance in response to wavelengths much greater than its size. The large material contrast is an essential prerequisite for the subwavelength resonant response.

In this review, we consider wave interaction with systems of subwavelength resonators. At particular low frequencies, known as subwavelength resonances, subwavelength resonators behave as strong wave scatterers. Using layer potential techniques and Gohberg-Sigal theory, we first derive a formula for the resonances of a system of resonators of arbitrary shapes. Then, we derive an effective medium theory for wave propagation in systems of resonators. We start with a multiple scattering formulation of the scattering problem in which an incident wave impinges on a large number of small, identical resonators in a homogeneous medium. Under certain conditions on the configuration of the resonator distribution, the point interaction approximation holds and yields an effective medium theory for the system of resonators as the number of resonators tends to infinity. As a consequence, below the resonant frequency of a single resonator, the obtained effective media may be highly refractive, making the focusing of waves at subwavelength scales achievable.

Then, we provide a mathematical theory for understanding the mechanism behind the double-negative refractive index phenomenon in systems of subwavelength resonators. The design of double-negative metamaterials generally requires the use of two different kinds of subwavelength resonators, which may limit the applicability of double-negative metamaterials. Herein we rely on media that consists of only a single type of resonant element, and show how to turn the metamaterial with a single negative effective property into a negative refractive index metamaterial, which acts as a superlens. Using dimers made of two identical resonators, we show that both effective material parameters can be negative near the anti-resonance of the two hybridized resonances for a single constituent dimer of subwavelength resonators.

Furthermore, we consider periodic structures of subwavelength resonators where subwavelength band gap opening typically occurs. This can induce rich physics on the subwavelength scale which cannot be understood by the standard homogenization theory. To demonstrate the opening of a subwavelength band gap, we exploit the strong interactions produced by subwavelength resonators among the cells in a periodic structure. We derive an approximate formula in terms of the contrast for the quasi-periodic subwavelength resonant frequencies of an arbitrarily shaped subwavelength resonator. Then, we consider the behavior of the first Bloch eigenfunction near the critical frequency where a subwavelength band gap of the periodic structure opens. For a square lattice, we show that the critical frequency occurs at the corner of the Brillouin zone where the Bloch eigenfunctions are antiperiodic. We develop a high-frequency homogenization technique to describe the rapid variations of the Bloch eigenfunctions on the microscale (the scale of the elementary crystal cell). Compared to the effective medium theory, an effective equation can be derived only for the envelope of this first Bloch eigenfunction.

Defect modes and guided modes can be shown to exist in perturbed subwavelength resonator crystals. We use the subwavelength band gap to demonstrate cavities and waveguides of subwavelength dimensions. First, by perturbing the size of a single resonator inside the crystal, we show that this crystal has a localized eigenmode close to the defect resonator. Further, by modifying the sizes of the subwavelength resonators along a line in a crystal, we show that the line defect acts as a waveguide; waves of certain frequencies will be localized to, and guided along, the line defect.

Topological properties of periodic lattices of subwavelength resonators are also considered, and we investigate the existence of Dirac cones in honeycomb lattices and topologically protected edge modes in chains of subwavelength resonators. We first show the existence of a Dirac dispersion cone in a honeycomb crystal comprised of subwavelength resonators of arbitrary shape. The high-frequency homogenization technique shows that, near the Dirac points, the Bloch eigenfunc-tion is the sum of two eigenmodes. Each eigenmode can be decomposed into two components: one which is slowly varying and satisfies a homogenized equation, while the other is periodic and highly oscillating. The slowly oscillating components of the eigenmodes satisfy a system of Dirac equations. This yields a near-zero effective refractive index near the Dirac points for the plane-wave envelopes of the Bloch eigenfunctions in a subwavelength metamaterial. The opening of a Dirac cone can create topologically protected edge modes, which are stable against geometric errors of the structure. We study a crystal which consists of a chain of subwavelength resonators arranged as dimers (often known as an SSH chain) and show that it exhibits a topologically non-trivial band gap, leading to robust localization properties at subwavelength scales.

Finally, we present a bio-inspired system of subwavelength resonators designed to mimic the cochlea. The system is inspired by the graded properties of the cochlear membranes, which are able to perform spatial frequency separation. Using layer potential techniques, the resonant modes of the system are computed and the model's ability to decompose incoming signals is explored.

This review is organized as follows. In Sect. 3.2, after stating the subwavelength resonance problem and introducing some preliminaries on the layer potential techniques and Gohberg-Sigal theory, we prove the existence of subwavelength resonances for systems of resonators and show a modal decomposition for the associated eigenmodes. Then, we study in Sect. 3.3 the behavior of the coupled subwavelength resonant modes when the subwavelength resonators are brought close together. In Sect. 3.4 we derive an effective medium theory for dilute systems of subwavelength resonators. Section 3.5 is devoted to the spectral analysis of periodic structures of subwavelength resonators. After recalling some preliminaries on the Floquet theory and quasi-periodic layer potentials, we prove the occurrence of subwavelength band gap opening in square lattices of subwavelength resonators and characterize the behavior of the first Bloch eigenfunction near the critical frequency where a subwavelength band gap of the periodic structure opens. In Sect. 3.6, we consider honeycomb lattices of subwavelength resonators and prove the existence of Dirac cones. We also study a chain of subwavelength resonators which exhibit a topologically non-trivial band gap. Finally, in Sect. 3.7, we present a graded array of subwavelength resonators which is designed to mimic the frequency separation proprieties of the cochlea. The review ends with some concluding remarks.

3.2 Subwavelength Resonances

We begin by describing the resonance problem and the main mathematical tools we will use to study a finite collection of subwavelength resonators.

3.2.1 Problem Setting

We are interested in studying wave propagation in a homogeneous background medium with $N \in \mathbb{N}$ disjoint bounded inclusions, which we label as

$$D_1, D_2, \ldots, D_N \subset \mathbb{R}^3.$$

We assume that the boundaries are all of class $C^{1,\eta}$ with $0 < \eta < 1$ and write $D = D_1 \cup \cdots \cup D_N$.

We denote the material parameters within the bounded regions D by ρ_b and κ_b, respectively. The corresponding parameters for the background medium are ρ and κ and the wave speeds in D and $\mathbb{R}^3 \setminus \overline{D}$ are given by $v_b = \sqrt{\kappa_b/\rho_b}$ and $v = \sqrt{\kappa/\rho}$. We define the wave numbers as

$$k = \frac{\omega}{v}, \quad k_b = \frac{\omega}{v_b}. \tag{3.1}$$

We also define the dimensionless contrast parameter

$$\delta = \frac{\rho_b}{\rho}. \tag{3.2}$$

We assume that

$$\delta \ll 1 \text{ while } v_b = \mathcal{O}(1) \text{ and } v = \mathcal{O}(1). \tag{3.3}$$

This high-contrast assumption will give the desired subwavelength behaviour, which we will study through an asymptotic analysis in terms of δ.

For $\omega \in \mathbb{C}$, we study the scattering problem

$$\begin{cases} \Delta u + k^2 u = 0 & \text{in } \mathbb{R}^3 \setminus \overline{D}, \\ \Delta u + k_b^2 u = 0 & \text{in } D, \\ u|_+ - u|_- = 0 & \text{on } \partial D, \\ \delta \dfrac{\partial u}{\partial v}\bigg|_+ - \dfrac{\partial u}{\partial v}\bigg|_- = 0 \text{ on } \partial D, \\ u(x) - u^{in}(x) & \text{satisfies the Sommerfeld radiation} \\ & \text{condition as } |x| \to \infty, \end{cases} \tag{3.4}$$

where $|_+$ and $|_-$ denote the limits from the outside and inside of D. Here, u^{in} is the incident field which we assume satisfies $\Delta u^{in} + k^2 u^{in} = 0$ in $\mathbb{R}^3$ and $\nabla u^{in}|_D = \mathcal{O}(\omega)$. We restrict to frequencies such that $\text{Re}(k) > 0$, whereby the Sommerfeld radiation condition is given by

$$\lim_{|x|\to\infty} |x| \left(\frac{\partial}{\partial |x|} - ik \right) u = 0, \tag{3.5}$$

which corresponds to the case where u radiates energy outwards (and not inwards).

Definition 3.2.1 (Subwavelength Resonant Frequency) We define a subwavelength resonant frequency to be $\omega = \omega(\delta)$ such that $\text{Re}(\omega) > 0$ and:

 (i) there exists a non-trivial solution to (3.4) when $u^{in} = 0$,
(ii) ω depends continuously on δ and is such that $\omega(\delta) \to 0$ as $\delta \to 0$.

The scattering problem (3.4) is a model problem for subwavelength resonators with high-contrast materials. It can be effectively studied using representations in terms of integral operators.

3.2.2   Layer Potential Theory on Bounded Domains and Gohberg-Sigal Theory

The layer potential operators are the main mathematical tools used in the study of the resonance problem described above. These are operator-valued holomorphic functions, and can be studied using Gohberg-Sigal theory.

Layer Potential Operators

Let $\mathcal{S}_D^k$ be the single layer potential, defined by

$$\mathcal{S}_D^k[\phi](x) := \int_{\partial D} G^k(x-y)\phi(y) \, d\sigma(y), \quad x \in \mathbb{R}^3, \tag{3.6}$$

where $G^k(x)$ is the outgoing Helmholtz Green's function, given by

$$G^k(x) := -\frac{e^{ik|x|}}{4\pi |x|}, \quad x \in \mathbb{R}^3, \ \text{Re}(k) \geq 0.$$

Here, "outgoing" refers to the fact that G^k satisfies the Sommerfeld radiation condition (3.5). For $k = 0$ we omit the superscript and write the fundamental solution to the Laplacian as G.

For the single layer potential corresponding to the Laplace equation, $\mathcal{S}_D^0$, we also omit the superscript and write $\mathcal{S}_D$. It is well known that the trace operator $\mathcal{S}_D : L^2(\partial D) \to H^1(\partial D)$ is invertible, where $H^1(\partial D)$ is the space of functions that are square integrable on ∂D and have a weak first derivative that is also square integrable. We denote by $\mathcal{L}(L^2(\partial D), H^1(\partial D))$ the set of bounded linear operators from $L^2(\partial D)$ into $H^1(\partial D)$.

The Neumann-Poincaré operator $\mathcal{K}_D^{k,*} : L^2(\partial D) \to L^2(\partial D)$ is defined by

$$\mathcal{K}_D^{k,*}[\phi](x) := \int_{\partial D} \frac{\partial}{\partial \nu_x} G^k(x - y)\phi(y) \, d\sigma(y), \quad x \in \partial D,$$

where $\partial/\partial \nu_x$ denotes the outward normal derivative at $x \in \partial D$. For $k = 0$ we omit the superscript and write $\mathcal{K}_D^*$.

The behaviour of $\mathcal{S}_D^k$ on the boundary ∂D is described by the following relations, often known as *jump relations*,

$$\mathcal{S}_D^k[\phi]\big|_+ = \mathcal{S}_D^k[\phi]\big|_-, \tag{3.7}$$

and

$$\frac{\partial}{\partial \nu} \mathcal{S}_D^k[\phi]\Big|_\pm = \left(\pm\frac{1}{2}I + \mathcal{K}_D^{k,*}\right)[\phi]. \tag{3.8}$$

When k is small, the single layer potential satisfies

$$\mathcal{S}_D^k = \mathcal{S}_D + k\mathcal{S}_{D,1} + k^2\mathcal{S}_{D,2} + k^3\mathcal{S}_{D,3} + \mathcal{O}(k^4), \tag{3.9}$$

where the error term is with respect to the operator norm $\|.\|_{\mathcal{L}(L^2(\partial D), H^1(\partial D))}$, and the operators $\mathcal{S}_{D,n} : L^2(\partial D) \to H^1(\partial D)$ for $n = 1, 2, 3$ are given by

$$\mathcal{S}_{D,n}[\phi](x) = -\frac{i^n}{4\pi n!} \int_{\partial D} |x - y|^{n-1}\phi(y) \, d\sigma(y) \quad x \in \partial D.$$

Moreover, we have

$$\mathcal{K}_D^{k,*} = \mathcal{K}_D^{0,*} + k^2\mathcal{K}_{D,2} + k^3\mathcal{K}_{D,3} + \mathcal{O}(k^4), \tag{3.10}$$

where the error term is with respect to the operator norm $\|.\|_{\mathcal{L}(L^2(\partial D), L^2(\partial D))}$ and where

$$\mathcal{K}_{D,2}[\phi](x) = \frac{1}{8\pi} \int_{\partial D} \frac{(x - y) \cdot \nu_x}{|x - y|}\phi(y) \, d\sigma(y),$$

and

$$\mathcal{K}_{D,3}[\phi](x) = \frac{i}{12\pi} \int_{\partial D} (x - y) \cdot \nu_x \phi(y) \, d\sigma(y),$$

for $x \in \partial D$. We have the following lemma from [11].

Lemma 3.2.1 *Let $N = 2$. For any $\varphi \in L^2(\partial D)$ we have, for $i = 1, 2$,*

$$\int_{\partial D_i} \left(-\frac{1}{2}I + \mathcal{K}_D^* \right) [\varphi] \, d\sigma = 0, \qquad \int_{\partial D_i} \left(\frac{1}{2}I + \mathcal{K}_D^* \right) [\varphi] \, d\sigma = \int_{\partial D_i} \varphi \, d\sigma,$$

$$\int_{\partial D_i} \mathcal{K}_{D,2}[\varphi] \, d\sigma = -\int_{D_i} \mathcal{S}_D[\varphi] \, dx, \qquad \int_{\partial D_i} \mathcal{K}_{D,3}[\varphi] \, d\sigma = \frac{i|D_i|}{4\pi} \int_{\partial D} \varphi \, d\sigma. \tag{3.11}$$

A thorough presentation of other properties of the layer potential operators and their use in wave-scattering problems can be found in e.g. [10].

Generalized Argument Principle and Generalized Rouché's Theorem

The Gohberg-Sigal theory refers to the generalization to operator-valued functions of two classical results in complex analysis, the argument principle and Rouché's theorem [10, 23, 24].

Let $\mathcal{B}$ and $\mathcal{B}'$ be two Banach spaces and denote by $\mathcal{L}(\mathcal{B}, \mathcal{B}')$ the space of bounded linear operators from $\mathcal{B}$ into $\mathcal{B}'$. A point z_0 is called a *characteristic value* of the operator-valued function $z \mapsto A(z) \in \mathcal{L}(\mathcal{B}, \mathcal{B}')$ if $A(z)$ is holomorphic in some neighborhood of z_0, except possibly at z_0 and there exists a vector-valued function $\phi(z)$ with values in $\mathcal{B}$ such that

(i) $\phi(z)$ is holomorphic at z_0 and $\phi(z_0) \neq 0$,
(ii) $A(z)\phi(z)$ is holomorphic at z_0 and vanishes at this point.

Let V be a simply connected bounded domain with rectifiable boundary ∂V. An operator-valued function $A(z)$ is normal with respect to ∂V if it is finitely meromorphic and of Fredholm type in V, continuous on ∂V, and invertible for all $z \in \partial V$.

If $A(z)$ is normal with respect to the contour ∂V and z_j, $j = 1, \ldots, \sigma$, are all its characteristic values and poles lying in V, the full multiplicity $\mathcal{M}(A; \partial V)$ of $A(z)$ for $z \in V$ is the number of characteristic values of $A(z)$ for $z \in V$, counted with their multiplicities, minus the number of poles of $A(z)$ in V, counted with their multiplicities:

$$\mathcal{M}(A; \partial V) := \sum_{j=1}^{\sigma} M(A(z_j)),$$

with $M(A(z_j))$ being the multiplicity of z_j; see [10, Chap. 1].

The following results are from [24].

Theorem 3.2.2 (Generalized Argument Principle) *Suppose that $A(z)$ is an operator-valued function which is normal with respect to ∂V. Let $f(z)$ be a scalar function which is holomorphic in V and continuous in $\overline{V}$. Then*

$$\frac{1}{2\pi i}\mathrm{tr}\int_{\partial V}f(z)A(z)^{-1}\frac{d}{dz}A(z)dz = \sum_{j=1}^{\sigma}M(A(z_j))f(z_j),$$

where z_j, $j = 1,\ldots,\sigma$, are all the points in V which are either poles or characteristic values of $A(z)$.

A generalization of Rouché's theorem to operator-valued functions is stated below.

Theorem 3.2.3 (Generalized Rouché's Theorem) *Let $A(z)$ be an operator-valued function which is normal with respect to ∂V. If an operator-valued function $S(z)$ which is finitely meromorphic in V and continuous on ∂V satisfies the condition*

$$\|A(z)^{-1}S(z)\|_{\mathcal{L}(\mathcal{B},\mathcal{B})} < 1, \quad z \in \partial V,$$

then $A + S$ is also normal with respect to ∂V and

$$\mathcal{M}(A; \partial V) = \mathcal{M}(A + S; \partial V).$$

3.2.3 Capacitance Matrix Analysis

The existence of subwavelength resonant frequencies is stated in the following theorem, which was proved in [1, 8] using Theorem 3.2.3.

Theorem 3.2.4 *A system of N subwavelength resonators exhibits N subwavelength resonant frequencies with positive real parts, up to multiplicity.*

Proof The solution u to the scattering problem (3.4) can be represented as

$$u(x) = \begin{cases} u^{in} + \mathcal{S}_D^k[\psi](x), & x \in \mathbb{R}^3 \setminus \overline{D}, \\ \mathcal{S}_D^{k_b}[\phi](x), & x \in D, \end{cases} \tag{3.12}$$

for some surface potentials $(\phi, \psi) \in L^2(\partial D) \times L^2(\partial D)$, which must be chosen so that u satisfies the transmission conditions across ∂D. Using the jump relation between $\mathcal{S}_D^k$ and $\mathcal{K}_D^{k,*}$, we see that in order to satisfy the transmission conditions on

∂D the densities ϕ and ψ must satisfy, for $x \in \partial D$,

$$\begin{cases} \mathcal{S}_D^{k_b}[\phi](x) - \mathcal{S}_D^k[\psi](x) = u^{in}(x), \\ \left(-\frac{1}{2}I + \mathcal{K}_D^{k_b,*}\right)[\phi](x) - \delta\left(\frac{1}{2}I + \mathcal{K}_D^{k,*}\right)[\psi](x) = \delta\frac{\partial u^{in}}{\partial \nu}(x). \end{cases} \tag{3.13}$$

Therefore, ϕ and ψ satisfy the following system of boundary integral equations:

$$\mathcal{A}(\omega, \delta)[\Psi] = F, \tag{3.14}$$

where

$$\mathcal{A}(\omega, \delta) = \begin{pmatrix} \mathcal{S}_D^{k_b} & -\mathcal{S}_D^k \\ -\frac{1}{2}I + (\mathcal{K}_D^{k_b})^* & -\delta(\frac{1}{2}I + (\mathcal{K}_D^k)^*) \end{pmatrix}, \quad \Psi = \begin{pmatrix} \phi \\ \psi \end{pmatrix}, \quad F = \begin{pmatrix} u^{in} \\ \delta\frac{\partial u^{in}}{\partial \nu} \end{pmatrix}. \tag{3.15}$$

One can show that the scattering problem (3.4) is equivalent to the system of boundary integral equations (3.14). It is clear that $\mathcal{A}(\omega, \delta)$ is a bounded linear operator from $\mathcal{H} := L^2(\partial D) \times L^2(\partial D)$ to $\mathcal{H}_1 := H^1(\partial D) \times L^2(\partial D)$. As defined in Theorem 3.2.1, the resonant frequencies to the scattering problem (3.4) are the complex numbers ω with positive imaginary part such that there exists a nontrivial solution to the following equation:

$$\mathcal{A}(\omega, \delta)[\Psi] = 0. \tag{3.16}$$

These can be viewed as the characteristic values of the holomorphic operator-valued function (with respect to ω) $\mathcal{A}(\omega, \delta)$. The subwavelength resonant frequencies lie in the right half of a small neighborhood of the origin in the complex plane. In what follows, we apply the Gohberg-Sigal theory to find these frequencies.

We first look at the limiting case when $\delta = \omega = 0$. It is clear that

$$\mathcal{A}_0 := \mathcal{A}(0, 0) = \begin{pmatrix} \mathcal{S}_D & -\mathcal{S}_D \\ -\frac{1}{2}I + \mathcal{K}_D^* & 0 \end{pmatrix}, \tag{3.17}$$

where $\mathcal{S}_D$ and $\mathcal{K}_D^*$ are respectively the single layer potential and the Neumann–Poincaré operator on ∂D associated with the Laplacian.

Since $\mathcal{S}_D : L^2(\partial D) \to H^1(\partial D)$ is invertible in dimension three and $\mathrm{Ker}(-\frac{1}{2}I + \mathcal{K}_D^*)$ has dimension equal to the number of connected components of D, it follows that $\mathrm{Ker}(\mathcal{A}_0)$ is of dimension N. This shows that $\omega = 0$ is a characteristic value for the holomorphic operator-valued function $\mathcal{A}(\omega, 0)$ of full multiplicity $2N$. By the generalized Rouché's theorem, we can conclude that for any δ, sufficiently small, there exist $2N$ characteristic values to the holomorphic operator-valued function $\mathcal{A}(\omega, \delta)$ such that $\omega_n(0) = 0$ and ω_n depends on δ continuously. N of these

characteristic values, $\omega_n = \omega_n(\delta), n = 1, \ldots, N$, have positive real parts, and these are precisely the subwavelength resonant frequencies of the scattering problem (3.4).
                                                                                            □

Our approach to approximate the subwavelength resonant frequencies is to study the *(weighted) capacitance matrix*, which offers a rigorous discrete approximation to the differential problem. The eigenstates of this $N \times N$-matrix characterise, at leading order in δ, the subwavelength resonant modes of the system of N resonators.

In order to introduce the notion of capacitance, we define the functions ψ_j, for $j = 1, \ldots, N$, as

$$\psi_j := \mathcal{S}_D^{-1}[\chi_{\partial D_j}], \tag{3.18}$$

where $\chi_A : \mathbb{R}^3 \to \{0, 1\}$ is used to denote the characteristic function of a set $A \subset \mathbb{R}^3$. The capacitance matrix $C = (C_{ij})$ is defined, for $i, j = 1, \ldots, N$, as

$$C_{ij} := -\int_{\partial D_i} \psi_j \, d\sigma. \tag{3.19}$$

In order to capture the behaviour of an asymmetric array of resonators we need to introduce the weighted capacitance matrix $C^{\mathrm{vol}} = (C_{ij}^{\mathrm{vol}})$, given by

$$C_{ij}^{\mathrm{vol}} := \frac{1}{|D_i|} C_{ij}, \tag{3.20}$$

which accounts for the differently sized resonators (see e.g. [11, 16, 17] for other variants in different settings).

We define the functions S_n^ω, for $n = 1 \ldots, N$, as

$$S_n^\omega(x) := \begin{cases} \mathcal{S}_D^k[\psi_n](x), & x \in \mathbb{R}^3 \setminus \overline{D}, \\ \mathcal{S}_D^{k_b}[\psi_n](x), & x \in D. \end{cases}$$

Lemma 3.2.2 *The solution to the scattering problem can be written, for $x \in \mathbb{R}^3$, as*

$$u(x) - u^{in}(x) = \sum_{n=1}^N q_n S_n^\omega(x) - \mathcal{S}_D^k \left[\mathcal{S}_D^{-1}[u^{in}] \right](x) + \mathcal{O}(\omega),$$

for coefficients $\underline{q} = (q_1, \ldots, q_N)$ which satisfy, up to an error of order $\mathcal{O}(\delta\omega + \omega^3)$,

$$\left(\omega^2 - v_b^2 \delta \, C^{\mathrm{vol}} \right) \underline{q} = v_b^2 \delta \begin{pmatrix} \frac{1}{|D_1|} \int_{\partial D_1} \mathcal{S}_D^{-1}[u^{in}] \, d\sigma \\ \vdots \\ \frac{1}{|D_N|} \int_{\partial D_N} \mathcal{S}_D^{-1}[u^{in}] \, d\sigma \end{pmatrix}.$$

Proof Using the asymptotic expansions (3.9) and (3.10) for $\mathcal{S}_D^k$ and $\mathcal{K}_D^{k,*}$ in (3.13), we can see that

$$\psi = \phi - \mathcal{S}_D^{-1}[u^{in}] + \mathcal{O}(\omega),$$

and, further, that

$$\left(-\frac{1}{2}I + \mathcal{K}_D^* + \frac{\omega^2}{v_b^2}\mathcal{K}_{D,2} - \delta\left(\frac{1}{2}I + \mathcal{K}_D^*\right) \right)[\phi] =$$

$$- \delta\left(\frac{1}{2}I + \mathcal{K}_D^*\right)\mathcal{S}_D^{-1}[u^{in}] + \mathcal{O}(\delta\omega + \omega^3). \quad (3.21)$$

At leading order, (3.21) says that $\left(-\frac{1}{2}I + \mathcal{K}_D^*\right)[\phi] = 0$ so, in light of the fact that $\{\psi_1, \ldots, \psi_N\}$ forms a basis for $\mathrm{Ker}\left(-\frac{1}{2}I + \mathcal{K}_D^*\right)$, the solution can be written as

$$\phi = \sum_{n=1}^{N} q_n\psi_n + \mathcal{O}(\omega^2 + \delta), \quad (3.22)$$

for coefficients $\underline{q} = (q_1, \ldots, q_N)$.

Finally, integrating (3.21) over ∂D_i, for $1 \leq i \leq N$, gives us that

$$-\omega^2 \int_{D_i} \mathcal{S}_D[\phi]\,\mathrm{d}x - v_b^2\delta \int_{\partial D_i} \phi\,\mathrm{d}\sigma = -v_b^2\delta \int_{\partial D_i} \mathcal{S}_D^{-1}[u^{in}]\,\mathrm{d}\sigma,$$

up to an error of order $\mathcal{O}(\delta\omega + \omega^3)$. Substituting the expression (3.22) gives the desired result. □

Theorem 3.2.5 *As* $\delta \to 0$, *the subwavelength resonant frequencies satisfy the asymptotic formula*

$$\omega_n = \sqrt{v_b^2\lambda_n\delta} - i\tau_n\delta + \mathcal{O}(\delta^{3/2}),$$

for $n = 1, \ldots, N$, *where* λ_n *are the eigenvalues of the weighted capacitance matrix* C^{vol} *and* τ_n *are non-negative real numbers that depend on* C, v *and* v_b.

Proof If $u^{in} = 0$, we find from Lemma 3.2.2 that there is a non-zero solution to the resonance problem when $\omega^2/v_b^2\delta$ is an eigenvalue of C^{vol}, at leading order.

To find the imaginary part, we adopt the ansatz

$$\omega_n = \sqrt{v_b^2\lambda_n\delta} - i\tau_n\delta + \mathcal{O}(\delta^{3/2}). \quad (3.23)$$

Using a few extra terms in the asymptotic expansions for $\mathcal{S}_D^k$ and $\mathcal{K}_D^{k,*}$, we have that

$$\psi = \phi + \frac{k_b - k}{4\pi \mathrm{i}} \left(\sum_{n=1}^N \psi_n \right) \int_{\partial D} \phi \, \mathrm{d}\sigma + \mathcal{O}(\omega^2),$$

and, hence, that

$$\left(-\frac{1}{2} I + \mathcal{K}_D^* + k_b^2 \mathcal{K}_{D,2} + k_b^3 \mathcal{K}_{D,3} - \delta \left(\frac{1}{2} I + \mathcal{K}_D^* \right) \right) [\phi]$$

$$- \frac{\delta(k_b - k)}{4\pi \mathrm{i}} \left(\sum_{n=1}^N \psi_n \right) \int_{\partial D} \phi \, \mathrm{d}\sigma = \mathcal{O}(\delta\omega^2 + \omega^4).$$

We then substitute the decomposition (3.22) and integrate over ∂D_i, for $i = 1, \ldots, N$, to find that, up to an error of order $\mathcal{O}(\delta\omega^2 + \omega^4)$, it holds that

$$\left(-\frac{\omega^2}{v_b^2} - \frac{\omega^3 \mathrm{i}}{4\pi v_b^3} J C + \delta C^{\mathrm{vol}} + \frac{\delta\omega \mathrm{i}}{4\pi} \left(\frac{1}{v_b} - \frac{1}{v} \right) C^{\mathrm{vol}} J C \right) \underline{q} = 0,$$

where J is the $N \times N$ matrix of ones (i.e. $J_{ij} = 1$ for all $i, j = 1, \ldots, N$). Then, using the ansatz (3.23) for ω_n we see that, if $\underline{v}_n$ is the eigenvector corresponding to λ_n, it holds that

$$\tau_n = \frac{v_b^2}{8\pi v} \frac{\underline{v}_n \cdot C J C \underline{v}_n}{\|\underline{v}_n\|_D^2}, \tag{3.24}$$

where we use the norm $\|x\|_D := \left(\sum_{i=1}^N |D_i| x_i^2 \right)^{1/2}$ for $x \in \mathbb{R}^N$. Since C is symmetric, we can see that $\tau_n \geq 0$. $\qquad\square$

It is more illustrative to rephrase Lemma 3.2.2 in terms of basis functions that are associated with the resonant frequencies. Denote by $\underline{v}_n = (v_{1,n}, \ldots, v_{N,n})$ the eigenvector of C^{vol} with eigenvalue λ_n. Then, we have a modal decomposition with coefficients that depend on the matrix $V = (v_{i,j})$, assuming the system is such that V is invertible. The following result follows from Lemma 3.2.2 by diagonalising the matrix C^{vol}.

Lemma 3.2.3 *Suppose that the resonators' geometry is such that the matrix of eigenvectors V is invertible. We define the functions*

$$u_n(x) = \sum_{i=1}^N v_{i,n} \, \mathcal{S}_D[\psi_i](x), \tag{3.25}$$

for n = 1, ..., N. Then if $\omega = \mathcal{O}(\sqrt{\delta})$ the solution to the scattering problem can be written, for $x \in \mathbb{R}^3$, as

$$u^s(x) := u(x) - u^{in}(x) = \sum_{n=1}^{N} a_n u_n(x) - \mathcal{S}_D \left[\mathcal{S}_D^{-1}[u^{in}] \right](x) + \mathcal{O}(\omega),$$

for coefficients which satisfy, up to an error of order $\mathcal{O}(\omega^3)$,

$$a_n(\omega^2 - \omega_n^2) = -A v_n \, \mathrm{Re}(\omega_n)^2,$$

where $v_n = \sum_{j=1}^{N} [V^{-1}]_{n,j}$, i.e. the sum of the nth row of V^{-1}.

Remark 3.2.1 When $N = 1$, the subwavelength resonant frequency ω_1 is called the Minnaert resonance. By writing an asymptotic expansion of $\mathcal{A}(\omega, \delta)$ in terms of δ and applying the generalized argument principle (Theorem 3.2.2), one can prove that ω_1 satisfies as δ goes to zero the asymptotic formula [8]

$$\omega_1 = \underbrace{\sqrt{\frac{\mathrm{Cap}_D}{|D|}} v_b \sqrt{\delta}}_{:= \omega_M} - \mathrm{i} \underbrace{\left(\frac{\mathrm{Cap}_D^2 v_b^2}{8\pi v |D|} \delta \right)}_{:= \tau_M} + \mathcal{O}(\delta^{\frac{3}{2}}), \tag{3.26}$$

where

$$\mathrm{Cap}_D := -\int_{\partial D} \mathcal{S}_D^{-1}[\chi_{\partial D}] \, d\sigma \tag{3.27}$$

is the capacity of ∂D. Moreover, the following monopole approximation of the scattered field for ω near ω_M holds [8]:

$$u^s(x) = g(\omega, \delta, D)(1 + o(1)) u^{in}(0) G^k(x), \tag{3.28}$$

with the origin $0 \in D$ and the scattering coefficient g being given by

$$g(\omega, \delta, D) = \frac{\mathrm{Cap}_D}{1 - (\frac{\omega_M}{\omega})^2 + \mathrm{i}\gamma_M}, \tag{3.29}$$

where the damping constant γ_M is given by

$$\gamma_M := \frac{(v + v_b)\mathrm{Cap}_D \omega}{8\pi v v_b} - \frac{(v - v_b)}{v} \frac{v_b \mathrm{Cap}_D^2 \delta}{8\pi |D| \omega}.$$

This shows the scattering enhancement near ω_M.

When $N = 2$, there are two subwavelength resonances with positive real part for the resonator dimer D. Assume that D_1 and D_2 are symmetric with respect to the origin 0 and let C_{ij}, for $i, j = 1, 2$, be defined by (3.19). Then, as $\delta \to 0$, by using Lemma 3.2.1 it follows that [11]

$$\omega_1 = \underbrace{\sqrt{(C_{11} + C_{12})v_b}\sqrt{\delta}}_{:=\omega_{M,1}} -i\tau_1\delta + \mathcal{O}(\delta^{3/2}), \tag{3.30}$$

$$\omega_2 = \underbrace{\sqrt{(C_{11} - C_{12})v_b}\sqrt{\delta}}_{:=\omega_{M,2}} +\delta^{3/2}\hat{\eta}_1 + i\delta^2\hat{\eta}_2 + \mathcal{O}(\delta^{5/2}), \tag{3.31}$$

where $\hat{\eta}_1$ and $\hat{\eta}_2$ are real numbers determined by D, v, and v_b and

$$\tau_1 = \frac{v_b^2}{4\pi v}(C_{11} + C_{12})^2.$$

The resonances ω_1 and ω_2 are called the hybridized resonances of the resonator dimmer D.

On the other hand, the resonator dimer can be approximated as a point scatterer with resonant monopole and resonant dipole modes. Assume that D_1 and D_2 are symmetric with respect to the origin. Then for $\omega = \mathcal{O}(\delta^{1/2})$ and $\delta \to 0$, and $|x|$ being sufficiently large, we have [11]

$$u^s(x) = \underbrace{g^0(\omega)u^{in}(0)G^k(x)}_{monopole}$$
$$+ \underbrace{\nabla u^{in}(0) \cdot g^1(\omega)\nabla G^k(x)}_{dipole} +\mathcal{O}(\delta|x|^{-1}), \tag{3.32}$$

where the scattering coefficients $g^0(\omega)$ and $g^1(\omega) = (g_{ij}^1(\omega))$ are given by

$$g^0(\omega) = \frac{C(1,1)}{1 - \omega_1^2/\omega^2}(1 + \mathcal{O}(\delta^{1/2})), \quad C(1,1) := C_{11} + C_{12} + C_{21} + C_{22}, \tag{3.33}$$

$$g_{ij}^1(\omega) = \int_{\partial D} \mathcal{S}_D^{-1}[x_i](y)y_j - \frac{\delta v_b^2}{\omega^2|D|(1 - \omega_2^2/\omega^2)}P^2\delta_{i1}\delta_{j1}, \tag{3.34}$$

with

$$P := \int_{\partial D} y_1(\psi_1 - \psi_2)(y) \, d\sigma(y), \tag{3.35}$$

ψ_i, for $i = 1, 2$, being defined by (3.18), and δ_{i1} and δ_{j1} being the Kronecker delta.

As shown in (3.28)–(3.29), around ω_M, a single resonator in free-space scatters waves with a greatly enhanced amplitude. If a second resonator is introduced, coupling interactions will occur giving according to (3.32) a system that has both monopole and dipole resonant modes. This pattern continues for larger number N of resonators [1].

Remark 3.2.2 The invertibility of V is a subtle issue and depends only on the geometry of the inclusions $D = D_1 \cup \cdots \cup D_N$. In the case that the resonators are all identical, V is clearly invertible since C^{vol} is symmetric.

Remark 3.2.3 In many cases $\tau_n = 0$ for some n (see for instance formula (3.31)), meaning the imaginary parts exhibit higher-order behaviour in δ. For example, the second (dipole) frequency for a pair of identical resonators is known to be $\mathcal{O}(\delta^2)$ [11]. In any case, the resonant frequencies will have negative imaginary parts, due to the radiation losses.

3.3 Close-to-Touching Subwavelength Resonators

In this section, we study the behaviour of the coupled subwavelength resonant modes when two subwavelength resonators are brought close together. We consider the case of a pair of spherical resonators and use bispherical coordinates to derive explicit representations for the capacitance coefficients which, as shown in Theorem 3.2.5, capture the system's resonant behaviour at leading order. We derive estimates for the rate at which the gradient of the scattered wave blows up as the resonators are brought together.

For simplicity, we study the effect of scattering by a pair of spherical inclusions, D_1 and D_2, with the same radius, which we denote by r, and separation distance ε (so that their centres are separated by $2r + \varepsilon$). We refer to [17] for the case of arbitrary sized spheres.

We choose the separation distance ε as a function of δ and will perform an asymptotic analysis in terms of δ. We choose ε to be such that, for some $0 < \beta < 1$,

$$\varepsilon \sim e^{-1/\delta^{1-\beta}} \text{ as } \delta \to 0. \tag{3.36}$$

As we will see shortly, with ε chosen to be in this regime the subwavelength resonant frequencies are both well behaved (*i.e.* $\omega = \omega(\delta) \to 0$ as $\delta \to 0$) and we can compute asymptotic expansions in terms of δ.

From Theorem 3.2.5 (see also Remark 3.2.1), there exist two subwavelength resonant modes, u_1 and u_2, with associated resonant frequencies ω_1 and ω_2 with positive real part, labelled such that $\text{Re}(\omega_1) < \text{Re}(\omega_2)$.

3.3.1 Coordinate System

The Helmholtz problem (3.4) is invariant under translations and rotations so we are free to choose the coordinate axes. Let R_j be the reflection with respect to ∂D_j and let p_1 and p_2 be the unique fixed points of the combined reflections $R_1 \circ R_2$ and $R_2 \circ R_1$, respectively. Let n be the unit vector in the direction of $p_2 - p_1$. We will make use of the Cartesian coordinate system (x_1, x_2, x_3) defined to be such that $p = (p_1 + p_2)/2$ is the origin and the x_3-axis is parallel to the unit vector n. Then one can see that [27]

$$p_1 = (0, 0, -\alpha) \quad \text{and} \quad p_2 = (0, 0, \alpha), \tag{3.37}$$

where

$$\alpha := \frac{\sqrt{\varepsilon(4r + \varepsilon)}}{2}. \tag{3.38}$$

Moreover, the sphere D_i is centered at $(0, 0, c_i)$ where

$$c_i = (-1)^i \sqrt{r^2 + \alpha^2}. \tag{3.39}$$

We then introduce a bispherical coordinate system (ξ, θ, φ) which is related to the Cartesian coordinate system (x_1, x_2, x_3) by

$$x_1 = \frac{\alpha \sin \theta \cos \varphi}{\cosh \xi - \cos \theta}, \quad x_2 = \frac{\alpha \sin \theta \sin \varphi}{\cosh \xi - \cos \theta}, \quad x_3 = \frac{\alpha \sinh \xi}{\cosh \xi - \cos \theta}, \tag{3.40}$$

and is chosen to satisfy $-\infty < \xi < \infty, 0 \le \theta < \pi$ and $0 \le \varphi < 2\pi$. The reason for this choice of coordinate system is that ∂D_1 and ∂D_2 are given by the level sets

$$\partial D_1 = \left\{ \xi = -\sinh^{-1}\left(\frac{\alpha}{r}\right) \right\}, \qquad \partial D_2 = \left\{ \xi = \sinh^{-1}\left(\frac{\alpha}{r}\right) \right\}. \tag{3.41}$$

This is depicted in Fig. 3.1 (for arbitrary sized spheres). The Cartesian coordinate system is chosen so that we can define a bispherical coordinate system (3.40) such that the boundaries of the two resonators are convenient level sets.

Fig. 3.1 Two close-to-touching spheres, annotated with the bispherical coordinate system outlined in Sect. 3.3.1

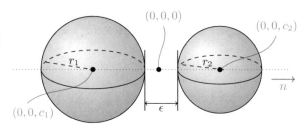

3.3.2   Resonant Frequency Hybridization and Gradient Blow-Up

Firstly, the resonant frequencies are given, in terms of the capacitance coefficients, by (see (3.30) and (3.31))

$$\omega_1 = \sqrt{\delta \frac{3v_b^2}{4\pi r^3}(C_{11} + C_{12})} + \mathcal{O}(\delta),$$

$$\omega_2 = \sqrt{\delta \frac{3v_b^2}{4\pi r^3}(C_{11} - C_{12})} + \mathcal{O}(\delta). \tag{3.42}$$

Further to this, the capacitance coefficients are given by

$$C_{11} = C_{22} = 8\pi\tilde{\alpha} \sum_{n=0}^{\infty} \frac{e^{(2n+1)\xi_0}}{e^{2(2n+1)\xi_0} - 1},$$

$$C_{12} = C_{21} = -8\pi\tilde{\alpha} \sum_{n=0}^{\infty} \frac{1}{e^{2(2n+1)\xi_0} - 1}, \tag{3.43}$$

where

$$\tilde{\alpha} := \sqrt{\varepsilon(r + \varepsilon/4)}, \qquad \xi_0 := \sinh^{-1}\left(\frac{\tilde{\alpha}}{r}\right).$$

From [30], we know the asymptotic behaviour of the series in (3.43) as $\xi_0 \to 0$, from which we can see that as $\varepsilon \to 0$,

$$C_{11} = 2\pi \frac{\tilde{\alpha}}{\xi_0}\left[\gamma + 2\ln 2 + \ln\left(\sqrt{r}\right) - \ln\left(\sqrt{\varepsilon}\right)\right] + \mathcal{O}(\varepsilon),$$

$$C_{12} = -2\pi \frac{\tilde{\alpha}}{\xi_0}\left[\gamma + \ln\left(\sqrt{r}\right) - \ln\left(\sqrt{\varepsilon}\right)\right] + \mathcal{O}(\varepsilon), \tag{3.44}$$

where $\gamma \approx 0.5772\ldots$ is the Euler–Mascheroni constant.

Combining (3.42) and (3.44) we reach the fact that the resonant frequencies are given, as $\delta \to 0$, by

$$\omega_1 = \sqrt{\delta \frac{3v_b^2 \ln 2}{r^2}} + \mathcal{O}(\delta),$$

$$\omega_2 = \sqrt{\delta \frac{3v_b^2}{2r^2}\left(\ln\left(\frac{r}{\varepsilon}\right) + 2\gamma + 2\ln 2\right)} + \mathcal{O}\left(\sqrt{\delta}\right). \tag{3.45}$$

Thus, the choice of $\varepsilon \sim e^{-1/\delta^{1-\beta}}$, where $0 < \beta < 1$, means that as $\delta \to 0$ we have that $\omega_1 \sim \sqrt{\delta}$ and $\omega_2 \sim \delta^{\beta/2}$.

The two resonant modes, u_1 and u_2, correspond to the two resonators oscillating in phase and in antiphase with one another, respectively. Since the eigenmode u_2 has different signs on the two resonators, ∇u_2 will blow up as the two resonators are brought together. Conversely, u_1 takes the same value on the two resonators so there will not be a singularity in the gradient. In particular, if we normalise the eigenmodes so that for any $x \in \partial D$

$$\lim_{\delta \to 0} |u_1(x)| \sim 1, \qquad \lim_{\delta \to 0} |u_2(x)| \sim 1, \tag{3.46}$$

then the choice of ε to satisfy the regime $\varepsilon \sim e^{-1/\delta^{1-\beta}}$ means that the maximal gradient of each eigenmode has the asymptotic behaviour, as $\delta \to 0$,

$$\max_{x \in \mathbb{R}^3 \setminus \overline{D}} |\nabla u_1(x)| \sim 1, \qquad \max_{x \in \mathbb{R}^3 \setminus \overline{D}} |\nabla u_2(x)| \sim \frac{1}{\varepsilon}. \tag{3.47}$$

By decomposing the scattered field into the two resonant modes, we can use (3.47) to understand the singular behaviour exhibited by the acoustic pressure. The solution u to the scattering problem (3.4) with incoming plane wave u^{in} with frequency $\omega = \mathcal{O}(\delta^{1/2})$ is given, for $x \in \mathbb{R}^3 \setminus \overline{D}$, by

$$u(x) = u^{in}(x) + a u_1(x) + b u_2(x), \tag{3.48}$$

where the coefficients a and b satisfy, as $\delta \to 0$, the equations

$$a(\omega^2 - \omega_1^2) = \frac{\delta v_b^2}{|D|} \int_{\partial D} \mathcal{S}_D^{-1}[u^{in}] \, d\sigma + \mathcal{O}(\delta^{\hat{\beta}}),$$

$$b(\omega^2 - \omega_2^2) = -\frac{\delta v_b^2}{|D|} \left(\int_{\partial D_1} \mathcal{S}_D^{-1}[u^{in}] \, d\sigma - \frac{|D_1|}{|D_2|} \int_{\partial D_2} \mathcal{S}_D^{-1}[u^{in}] \, d\sigma \right) + \mathcal{O}(\delta^{\hat{\beta}}),$$

with $\hat{\beta} := \min(2 - \beta, 3/2)$ and $|D|$ being the volume of $D = D_1 \cup D_2$.

3.4 Effective Medium Theory for Subwavelength Resonators

3.4.1 High Refractive Index Effective Media

We consider a domain Ω which contains a large number of small, identical resonators. If D_0 is a fixed domain, then for some $r > 0$ the N resonators are given, for $1 \leq j \leq N$, by

$$D_{0,j}^{r,N} = r D_0 + z_j^N,$$

for positions z_j^N. We always assume that r is sufficiently small such that the resonators are not overlapping and that $D_0^{r,N} = \bigcup_{j=1}^N D_{0,j}^{r,N} \Subset \Omega$.

We find the effective equation in the specific case that the frequency $\omega = \mathcal{O}(1)$ and satisfies

$$1 - (\frac{\omega_M}{\omega})^2 = \beta_0 r^{\varepsilon_0}, \tag{3.49}$$

for some fixed $0 < \varepsilon_0 < 1$ and constant β_0. We note that there are two cases depending on whether $\omega > \omega_M$ or $\omega < \omega_M$. In the former case, $\beta_0 > 0$ while in the latter case we have $\beta_0 < 0$.

Moreover, we assume that there exists some positive number Λ independent of N such that

$$r^{1-\varepsilon_0} N = \Lambda \quad \text{and } \Lambda \text{ is large.} \tag{3.50}$$

Since the resonators are small, we can use the point-scatter approximation from (3.28) to describe how they interact with incoming waves. To do so, we must make some extra assumptions on the regularity of the distribution $\{z_j^N : 1 \leq j \leq N\}$ so that the system is well behaved as $N \to \infty$ (under the assumption (3.50)). In particular, we assume that there exists some constant η such that for any N it holds that

$$\min_{i \neq j} |z_i^N - z_j^N| \geq \frac{\eta}{N^{1/3}}, \tag{3.51}$$

and, further, there exists some $0 < \varepsilon_1 < 1$ and constants $C_1, C_2 > 0$ such that for all $h \geq 2\eta N^{-1/3}$,

$$\sum_{|x-z_j^N| \geq h} \frac{1}{|x - y_j^N|^2} \leq C_1 N |h|^{-\varepsilon_1}, \qquad \text{uniformly for all } x \in \Omega, \tag{3.52}$$

$$\sum_{2\eta N^{-1/3} \leq |x-z_j^N| \leq 3h} \frac{1}{|x - y_j^N|} \leq C_2 N |h|, \qquad \text{uniformly for all } x \in \Omega. \tag{3.53}$$

Finally, we also need that

$$\varepsilon_2 := \frac{\varepsilon_0}{1 - \varepsilon_0} - \frac{\varepsilon_1}{3} > 0. \tag{3.54}$$

If we represent the field that is scattered by the collection of resonators

$$D_0^{r,N} = \bigcup_{j=1}^{N} D_{0,j}^{r,N}$$

as

$$u^N(x) = \begin{cases} u^{in}(x) + \mathcal{S}_{D_0^{r,N}}^k [\psi^N](x), & x \in \mathbb{R}^3 \setminus \overline{D_0^{r,N}}, \\ \mathcal{S}_{D_0^{r,N}}^{k_0} [\phi^N](x), & x \in D_0^{r,N}, \end{cases}$$

for some $\psi^N, \phi^N \in L^2(\partial D_0^{r,N})$, then we have the following lemma, which follows from [6, Proposition 3.1]. This justifies using a point-scatter approximation to describe the total incident field acting on the resonator $D_{0,j}^{r,N}$ and the scattered field due to $D_{0,j}^{r,N}$, defined respectively as

$$u_j^{in,N} = u^{in} + \sum_{i \neq j} \mathcal{S}_{D_{0,i}^{r,N}}^k [\psi^N] \quad \text{and} \quad u_j^{s,N} = \mathcal{S}_{D_{0,j}^{r,N}}^k [\psi^N].$$

Lemma 3.4.1 *Under the assumptions (3.49)–(3.54), it holds that the total incident field acting on the resonator $D_{0,j}^{r,N}$ is given, at z_j^N, by*

$$u_j^{in,N}(z_j^N) = u^{in}(z_j^N) + \sum_{i \neq j} \frac{r \text{Cap}_{D_0}}{1 - (\frac{\omega_M}{\omega})^2} G^k(z_j^N - z_i^N) u^{in}(z_j^N),$$

up to an error of order $\mathcal{O}(N^{-\varepsilon_2})$. Similarly, it holds that the scattered field due to the resonator $D_{0,j}^{r,N}$ is given, at x such that $|x - z_j^N| \gg r$, by

$$u_j^{s,N}(x) = \frac{r \text{Cap}_{D_0}}{1 - (\frac{\omega_M}{\omega})^2} G^k(x - z_j^N) u_j^{in,N}(z_j^N),$$

up to an error of order $\mathcal{O}(N^{-\varepsilon_2} + r|x - z_j^N|^{-1})$.

In order for the sums in Lemma 3.4.1 to be well behaved as $N \to \infty$, we make one additional assumption on the regularity of the distribution: that there exists a real-valued function $\tilde{V} \in C^1(\overline{\Omega})$ such that for any $f \in C^{0,\alpha}(\Omega)$, with $0 < \alpha \leq 1$, there is a constant C_3 such that

$$\max_{1 \leq j \leq N} \left| \frac{1}{N} \sum_{i \neq j} G^k(z_j^N - z_i^N) f(z_i^N) - \int_\Omega G^k(z_j^N - y) \tilde{V}(y) f(y) \, dy \right|$$
$$\leq C_3 \frac{1}{N^{\alpha/3}} \|f\|_{C^{0,\alpha}(\Omega)}. \tag{3.55}$$

Remark 3.4.1 It holds that $\widetilde{V} \geq 0$. If the resonators' centres $\{z_j^N : j = 1, \ldots, N\}$ are uniformly distributed, then $\widetilde{V}$ will be a positive constant, $\widetilde{V} = 1/|\Omega|$.

Under all these assumptions, we are able to derive effective equations for the system with an arbitrarily large number of small resonators. If we let $\varepsilon_3 \in (0, \frac{1}{3})$, then we will seek effective equations on the set given by

$$Y_{\varepsilon_3}^N := \left\{ x \in \mathbb{R}^3 : |x - z_j^N| \geq N^{\varepsilon_3 - 1} \text{ for all } 1 \leq j \leq N \right\}, \tag{3.56}$$

which is the set of points that are sufficiently far from the resonators, avoiding the singularities of the Green's function. The following result from [6] holds.

Theorem 3.4.1 *Let $\omega < \omega_M$. Under the assumptions* (3.50)–(3.55), *the solution u^N to the scattering problem* (3.4) *with the system of resonators $D_0^{r,N} = \bigcup_{j=1}^N D_{0,j}^{r,N}$ converges to the solution of*

$$\begin{cases} \left(\Delta + k^2 - \frac{\Lambda \mathrm{Cap}_{D_0}}{\beta_0} \widetilde{V}(x) \right) u(x) = 0, & x \in \Omega, \\ \left(\Delta + k^2 \right) u(x) = 0, & x \in \mathbb{R}^3 \setminus \overline{\Omega}, \\ u|_+ = u|_- & \text{on } \partial\Omega, \end{cases}$$

as $N \to \infty$, together with a radiation condition governing the behaviour in the far field, which says that uniformly for all $x \in Y_{\varepsilon_3}^N$ it holds that

$$|u^N(x) - u(x)| \leq CN^{-\min\left\{ \frac{1 - \varepsilon_0}{6}, \varepsilon_2, \varepsilon_3, \frac{1 - \varepsilon_3}{3} \right\}}.$$

By our assumption, $k = \mathcal{O}(1)$, $\widetilde{V} = \mathcal{O}(1)$, and $\beta_0 < 0$. When

$$-\Lambda \mathrm{Cap}_{D_0}/\beta_0 \gg 1,$$

we see that we have an effective high refractive index medium. As a consequence, this together with [5] gives a rigorous mathematical theory for the super-focusing experiment in [29]. Similarly to Theorem 3.4.1, if $\omega > \omega_M$, to the scattering problem (3.4) with the system of resonators $D_0^{r,N}$ converges to the solution of the following dissipative equation

$$\begin{cases} \left(\Delta + k^2 - \frac{\Lambda \mathrm{Cap}_{D_0}}{\beta_0} \widetilde{V}(x) \right) u(x) = 0, & x \in \Omega, \\ \left(\Delta + k^2 \right) u(x) = 0, & x \in \mathbb{R}^3 \setminus \overline{\Omega}, \\ u|_+ = u|_- & \text{on } \partial\Omega, \end{cases}$$

x as $N \to \infty$, together with a radiation condition governing the behaviour in the far field, which says that uniformly for all $x \in Y_{\varepsilon_3}^N$ it holds that

$$|u^N(x) - u(x)| \leq CN^{-\min\left\{\frac{1-\varepsilon_0}{6}, \varepsilon_2, \varepsilon_3, \frac{1-\varepsilon_3}{3}\right\}}.$$

Remark 3.4.2 At the resonant frequency $\omega = \omega_M$, the scattering coefficient g defined by (3.29) is of order one. Thus each scatterer is a point source with magnitude one. As a consequence, the addition or removal of one resonator from the medium affects the total field by a magnitude of the same order as the incident field. Therefore, we cannot expect any effective medium theory for the medium at this resonant frequency.

3.4.2 Double-Negative Metamaterials

In this subsection, we show that, using dimers of identical subwavelength resonators, the effective material parameters of dilute system of dimers can both be negative over a non empty range of frequencies [11]. As shown in (3.32), a dimer of identical subwavelength resonators can be approximated as a point scatterer with monopole and dipole modes. As seen before, it features two slightly different subwavelength resonances, called the hybridized resonances. The hybridized resonances are fundamentally different modes. The first mode is, as in the case of a single resonator, a monopole mode, while the second mode is a dipole mode. The resonance associated with the dipole mode is usually referred to as the anti-resonance.

For an appropriate volume fraction, when the excitation frequency is close to the anti-resonance, a double-negative effective ρ and κ for media consisting of a large number of dimers with certain conditions on their distribution can be obtained. The dipole modes in the background medium contribute to the effective ρ while the monopole modes contribute to the effective κ.

Here we consider the scattering of an incident plane wave u^{in} by N identical dimers with different orientations distributed in a homogeneous medium in $\mathbb{R}^3$. The N identical dimers are generated by scaling the normalized dimer D by a factor r, and then rotating the orientation and translating the center. More precisely, the dimers occupy the domain

$$D^N := \cup_{1 \leq j \leq N} D_j^N,$$

where $D_j^N = z_j^N + rR_{d_j^N}D$ for $1 \leq j \leq N$, with z_j^N being the center of the dimer D_j^N, r being the characteristic size, and $R_{d_j^N}$ being the rotation in $\mathbb{R}^3$ which aligns the dimer D_j^N in the direction d_j^N. Here, d_j^N is a vector of unit length in $\mathbb{R}^3$. For simplicity, we suppose that D is made of two identical spherical resonators. We also

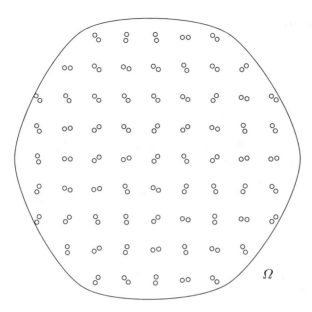

Fig. 3.2 Illustration of the dilute system of subwavelength dimers

assume that $0 < r \ll 1$, $N \gg 1$ and that $\{z_j^N : 1 \le j \le N\} \subset \Omega$ where Ω is a bounded domain. See Fig. 3.2.

The scattering of waves by the dimers can be modeled by the following system of equations:

$$
\begin{cases}
\nabla \cdot \dfrac{1}{\rho}\nabla u^N + \dfrac{\omega^2}{\kappa} u^N = 0 & \text{in } \mathbb{R}^3 \backslash \overline{D^N}, \\[2mm]
\nabla \cdot \dfrac{1}{\rho_b}\nabla u^N + \dfrac{\omega^2}{\kappa_b} u^N = 0 & \text{in } D^N, \\[2mm]
u_+^N - u_-^N = 0 & \text{on } \partial D^N, \\[2mm]
\dfrac{1}{\rho}\dfrac{\partial u^N}{\partial \nu}\bigg|_+ - \dfrac{1}{\rho_b}\dfrac{\partial u^N}{\partial \nu}\bigg|_- = 0 & \text{on } \partial D^N, \\[2mm]
u^N - u^{in} \text{ satisfies the Sommerfeld radiation condition,}
\end{cases}
\tag{3.57}
$$

where u^N is the total field.

We make the following assumptions:

(i) $\delta = \mu^2 r^2$ for some positive number $\mu > 0$;

(ii) $\omega = \omega_{M,2} + a r^2$ for some real number $a < \mu^3 \hat{\eta}_1$, where $\omega_{M,2}$ is defined in (3.31);

(iii) $rN = \Lambda$ for some positive number $\Lambda > 0$;

(iv) The dimers are regularly distributed in the sense that

$$\min_{i \neq j} |z_i^N - z_j^N| \geq \eta N^{-\frac{1}{3}},$$

for some constant η independent of N;

(v) There exists a function $\widetilde{V} \in C^1(\bar{\Omega})$ such that for any $f \in C^{0,\alpha}(\Omega)$ with $0 < \alpha \leq 1$, (3.55) holds for some constant C independent of N;

(vi) There exists a matrix valued function $\widetilde{B} \in C^1(\bar{\Omega})$ such that for $f \in (C^{0,\alpha}(\Omega))^3$ with $0 < \alpha \leq 1$,

$$\max_{1 \leq j \leq N} |\frac{1}{N} \sum_{i \neq j} (f(z_i^N) \cdot d_i^N)(d_i^N \cdot \nabla G^k(z_i^N - z_j^N)) - \int_\Omega f(y) \widetilde{B} \nabla_y G^k(y - z_j^N) \, dy|$$

$$\leq C \frac{1}{N^{\frac{\alpha}{3}}} \|f\|_{C^{0,\alpha}(\Omega)}$$

for some constant C independent of N;

(vii) There exists a constant $C > 0$ such that

$$\max_{1 \leq j \leq N} \frac{1}{N} \sum_{i \neq j} \frac{1}{|z_j^N - z_i^N|} \leq C, \qquad \max_{1 \leq j \leq N} \frac{1}{N} \sum_{i \neq j} \frac{1}{|z_j^N - z_i^N|^2} \leq C,$$

for all $1 \leq j \leq N$.

We introduce the two constants

$$\widetilde{g}^0 = \frac{2(C_{11} + C_{12})}{1 - \omega_{M,1}^2/\omega_{M,2}^2}, \qquad \widetilde{g}^1 = \frac{\mu^2 v_b^2}{2|D|\omega_{M,2}(\mu^3 \hat{\eta}_1 - a)} P^2,$$

where P is defined by (3.35), $\omega_{M,1}$ and $\omega_{M,2}$ are the leading orders in the asymptotic expansions (3.30) and (3.31) of the hybridized resonant frequencies as $\delta \to 0$, and the two functions

$$M_1 = \begin{cases} I & \text{in } \mathbb{R}^3 \setminus \Omega, \\ I - \Lambda \widetilde{g}^1 \widetilde{B} & \text{in } \Omega, \end{cases}$$

and

$$M_2 = \begin{cases} k^2 & \text{in } \mathbb{R}^3 \setminus \Omega, \\ k^2 - \Lambda \widetilde{g}^0 \widetilde{V} & \text{in } \Omega. \end{cases}$$

The following result from [11] holds.

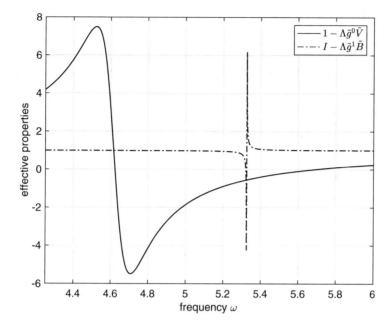

Fig. 3.3 Effective properties of the homogenized medium

Theorem 3.4.2 *Suppose that there exists a unique solution u to*

$$\nabla \cdot M_1(x)\nabla u(x) + M_2(x)u(x) = 0 \quad in \; \mathbb{R}^3, \tag{3.58}$$

such that $u - u^{in}$ satisfies the Sommerfeld radiation condition at infinity. Then, under assumptions (i)–(vii), we have $u^N(x) \to u(x)$ uniformly for $x \in \mathbb{R}^3$ such that $|x - z_j^N| \gg N^{-1}$ for all $1 \leq j \leq N$.

Note that from (3.43), it follows that $\omega_{M,2} > \omega_{M,1}$. Therefore, for large Λ, both the matrix M_1 and the scalar function M_2 are negative in Ω. See Fig. 3.3.

3.5 Periodic Structures of Subwavelength Resonators

In this section we investigate whether there is a possibility of subwavelength band gap opening in subwavelength resonator crystals. We first formulate the spectral problem for a subwavelength resonator crystal. Then we derive an asymptotic formula for the quasi-periodic resonances in terms of the contrast between the densities outside and inside the resonators. We prove the existence of a subwavelength band gap and estimate its width.

3.5.1   Floquet Theory

Let $f(x)$ for $x \in \mathbb{R}^d, d = 1, 2, 3$, be a function decaying sufficiently fast. We let $l_1, \ldots, l_d$ be linearly independent lattice vectors, and define the unit cell Y and the lattice Λ as

$$Y = \left\{ \sum_{n=1}^{d} s_n l_n \,\bigg|\, 0 < s_n < 1 \right\}, \qquad \Lambda = \left\{ \sum_{n=1}^{d} q_n l_n \,\bigg|\, q_n \in \mathbb{N} \right\}.$$

The Floquet transform of f is defined as:

$$\mathcal{U}[f](x, \alpha) = \sum_{n \in \Lambda} f(x - n) e^{i\alpha \cdot n}. \tag{3.59}$$

This transform is an analogue of the Fourier transform for the periodic case. The parameter α is called the quasi-periodicity, and it is an analogue of the dual variable in the Fourier transform. If we shift x by a period $m \in \Lambda$, we get the Floquet condition (or quasi-periodic condition)

$$\mathcal{U}[f](x + m, \alpha) = e^{i\alpha \cdot m} \mathcal{U}[f](x, \alpha), \tag{3.60}$$

which shows that it suffices to know the function $\mathcal{U}[f](x, \alpha)$ on the unit cell Y in order to recover it completely as a function of the x-variable. Moreover, $\mathcal{U}[f](x, \alpha)$ is periodic with respect to α:

$$\mathcal{U}[f](x, \alpha + q) = \mathcal{U}[f](x, \alpha), \quad q \in \Lambda^*. \tag{3.61}$$

Here, Λ^* is the dual lattice, generated by the dual lattice vectors $\alpha_1, \ldots, \alpha_d$ defined through the relation

$$l_i \alpha_j = 2\pi \delta_{ij}, \qquad 0 \le i, j \le d.$$

Therefore, α can be considered as an element of the torus $\mathbb{R}^d / \Lambda^*$. Another way of saying this is that all information about $\mathcal{U}[f](x, \alpha)$ is contained in its values for α in the fundamental domain Y^* of the dual lattice Λ^*. This domain is referred to as the Brillouin zone and is depicted in Fig. 3.4 for a square lattice in two dimensions.

The following result is an analogue of the Plancherel theorem when one uses the Fourier transform. Suppose that the measures $d\alpha$ and the Brillouin zone Y^* are normalized.

The following theorem holds [28].

Theorem 3.5.1 (Plancherel-Type Theorem) *The transform*

$$\mathcal{U} : L^2(\mathbb{R}^d) \to L^2(Y^*, L^2(Y))$$

Fig. 3.4 (First) Brillouin
zone for a square lattice in
two dimensions, with the
symmetry points Γ, X and
M. The highlighted triangle is
known as the reduced
Brillouin zone

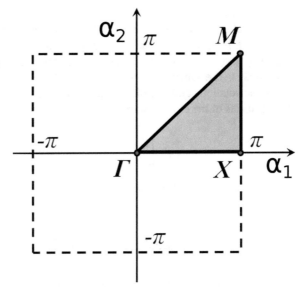

is isometric. Its inverse is given by

$$\mathcal{U}^{-1}[g](x) = \int_{Y^*} g(x, \alpha)\, d\alpha,$$

where the function $g(x, \alpha) \in L^2(Y \times Y^*)$ is extended from Y to all $x \in \mathbb{R}^d$ according to the Floquet condition (3.60).

Consider now a linear partial differential operator $L(x, \partial_x)$, whose coefficients are periodic with respect to Λ. Due to periodicity, the operator commutes with the Floquet transform

$$\mathcal{U}[Lf](x, \alpha) = L(x, \partial_x)\mathcal{U}[f](x, \alpha).$$

For each α, the operator $L(x, \partial_x)$ now acts on functions satisfying the corresponding Floquet condition (3.60). Denoting this operator by $L(\alpha)$, we see that the Floquet transform $\mathcal{U}$ expands the periodic partial differential operator L in $L^2(\mathbb{R}^d)$ into the direct integral of operators

$$\int_{Y^*}^{\oplus} L(\alpha)\, d\alpha. \tag{3.62}$$

If L is a self-adjoint operator, one can prove the main spectral result:

$$\sigma(L) = \bigcup_{\alpha \in Y^*} \sigma(L(\alpha)), \tag{3.63}$$

where σ denotes the spectrum.

If L is elliptic, the operators $L(\alpha)$ have compact resolvents and hence discrete spectra. If L is bounded from below, the spectrum of $L(\alpha)$ accumulates only at $+\infty$. Denote by $\mu_n(\alpha)$ the nth eigenvalue of $L(\alpha)$ (counted in increasing order with their multiplicity). The function $\alpha \mapsto \mu_n(\alpha)$ is continuous in Y^*. It is one branch of the dispersion relations and is called a *band function*. We conclude that the spectrum $\sigma(L)$ consists of the closed intervals (called the spectral bands)

$$\left[\min_\alpha \mu_n(\alpha), \max_\alpha \mu_n(\alpha) \right],$$

where $\min_\alpha \mu_n(\alpha) \to +\infty$ when $n \to +\infty$.

3.5.2 Quasi-Periodic Layer Potentials

We introduce a quasi-periodic version of the layer potentials. Again, we let Y and Y^* be the unit cell and dual unit cell, respectively. Let δ be the Dirac delta function. For $\alpha \in Y^*$, the function $G^{\alpha,k}$ is defined to satisfy

$$(\Delta_x + k^2)G^{\alpha,k}(x, y) = \sum_{m \in \Lambda} \delta(x - y - n)e^{im\cdot\alpha},$$

where $G^{\alpha,k}$ is α-quasi-periodic, i.e., $e^{-i\alpha\cdot x}G^{\alpha,k}(x, y)$ is periodic in x with respect to Y. It is known that $G^{\alpha,k}$ can be written as

$$G^{\alpha,k}(x, y) = \sum_{q \in \Lambda^*} \frac{e^{i(\alpha+q)\cdot(x-y)}}{k^2 - |\alpha + q|^2},$$

if $k \neq |\alpha + q|$ for any $q \in Y^*$. We remark that

$$G^{\alpha,k}(x, y) = G^{\alpha,0}(x, y) - G_l^{\alpha,\#}(x - y) := G^{\alpha,0}(x, y) - \sum_{l=1}^{\infty} k^{2l} \sum_{q \in \Lambda^*} \frac{e^{i(\alpha+q)\cdot(x-y)}}{|\alpha + q|^{2(l+1)}}$$

$$(3.64)$$

when $\alpha \neq 0$, and $k \to 0$.

We let D be as in Sect. 3.5.4 and additionally assume $D \Subset Y$. Then the quasi-periodic single layer potential $S_D^{\alpha,k}$ is defined by

$$S_D^{\alpha,k}[\phi](x) = \int_{\partial D} G^{\alpha,k}(x, y)\phi(y)\,d\sigma(y), \quad x \in \mathbb{R}^3. \tag{3.65}$$

It satisfies the following jump formulas:

$$\mathcal{S}_D^{\alpha,k}[\phi]\big|_+ = \mathcal{S}_D^{\alpha,k}[\phi]\big|_-,$$

and

$$\frac{\partial}{\partial \nu} \mathcal{S}_D^{\alpha,k}[\phi]\big|_\pm = \left(\pm \frac{1}{2}I + (\mathcal{K}_D^{-\alpha,k})^*\right)[\phi] \quad \text{on } \partial D,$$

where $(\mathcal{K}_D^{-\alpha,k})^*$ is the operator given by

$$(\mathcal{K}_D^{-\alpha,k})^*[\phi](x) = \int_{\partial D} \frac{\partial}{\partial \nu_x} G^{\alpha,k}(x,y)\phi(y)\, d\sigma(y).$$

We remark that $\mathcal{S}_D^{\alpha,0} : L^2(\partial D) \to H^1(\partial D)$ is invertible for $\alpha \neq 0$ [10]. Moreover, the following decomposition holds for the layer potential $\mathcal{S}_D^{\alpha,k}$:

$$\mathcal{S}_D^{\alpha,k} = \mathcal{S}_D^{\alpha,0} + k^2 \mathcal{S}_{D,1}^\alpha + \mathcal{O}(k^4) \quad \text{with} \quad \mathcal{S}_{D,1}^\alpha[\psi] := \int_{\partial D} G_1^{\alpha,\#}(x-y)\psi(y)\, d\sigma(y),$$

$$(3.66)$$

where the error term is with respect to the operator norm $\|\cdot\|_{\mathcal{L}(L^2(\partial D), H^1(\partial D))}$. Furthermore, analogously to (3.10), we have

$$(\mathcal{K}_D^{-\alpha,k})^* = (\mathcal{K}_D^{-\alpha,k})^* + k^2 \mathcal{K}_{D,1}^\alpha + \mathcal{O}(k^3), \tag{3.67}$$

where the error term is with respect to the operator norm $\|\cdot\|_{\mathcal{L}(L^2(\partial D), L^2(\partial D))}$.
    Finally, we introduce the α-quasi capacity of D, denoted by $\mathrm{Cap}_{D,\alpha}$,

$$\mathrm{Cap}_{D,\alpha} := \int_{Y\setminus\overline{D}} |\nabla u|^2 \, dy,$$

where u is the α-quasi-periodic harmonic function in $Y \setminus \overline{D}$ with $u = 1$ on ∂D. For $\alpha \neq 0$, we have $u(x) = \mathcal{S}_D^{\alpha,0}\left(\mathcal{S}_D^{\alpha,0}\right)^{-1}[\chi_{\partial D}](x)$ for $x \in Y \setminus \overline{D}$ and

$$\mathrm{Cap}_{D,\alpha} := -\int_{\partial D} \left(\mathcal{S}_D^{\alpha,0}\right)^{-1}[\chi_{\partial D}](y)\, d\sigma(y). \tag{3.68}$$

Moreover, we have a variational definition of $\mathrm{Cap}_{D,\alpha}$. Indeed, let $\mathcal{C}_\alpha^\infty(Y)$ be the set of $\mathcal{C}^\infty$ functions in Y which can be extended to $\mathcal{C}^\infty$ α-quasi-periodic functions in $\mathbb{R}^3$. Let $\mathcal{H}_\alpha$ be the closure of the set $\mathcal{C}_\alpha^\infty(Y)$ in $H^1(Y)$, and let $\mathcal{V}_\alpha := \{v \in \mathcal{H}_\alpha : v =$

1 on ∂D}. Then we can show that

$$\text{Cap}_{D,\alpha} = \min_{v \in \mathcal{V}_\alpha} \int_{Y \backslash \overline{D}} |\nabla v|^2 \, dy. \tag{3.69}$$

3.5.3 Square Lattice Subwavelength Resonator Crystal

We first describe the crystal under consideration. Assume that the resonators occupy $\cup_{n \in \mathbb{Z}^d} (D + n)$ for a bounded and simply connected domain $D \Subset Y$ with $\partial D \in C^{1,\eta}$ with $0 < \eta < 1$. See Fig. 3.5. As before, we denote by ρ_b and κ_b the material parameters inside the resonators and by ρ and κ the corresponding parameters for the background media and let $v, v_b, k,$ and k_b be defined by (3.1). We also let the dimensionless contrast parameter δ be defined by (3.2) and assume for simplicity that $v_b/v = 1$.

To investigate the phononic gap of the crystal we consider the following α–quasi-periodic equation in the unit cell $Y = [-1/2, 1/2)^3$:

$$\begin{cases} \nabla \cdot \dfrac{1}{\rho} \nabla u + \dfrac{\omega^2}{\kappa} u = 0 & \text{in} \quad Y \backslash \overline{D}, \\ \nabla \cdot \dfrac{1}{\rho_b} \nabla u + \dfrac{\omega^2}{\kappa_b} u = 0 & \text{in} \quad D, \\ u|_+ - u|_- = 0 & \text{on} \quad \partial D, \\ \dfrac{1}{\rho} \dfrac{\partial u}{\partial v}\bigg|_+ - \dfrac{1}{\rho_b} \dfrac{\partial u}{\partial v}\bigg|_- = 0 & \text{on} \quad \partial D, \\ e^{-i\alpha \cdot x} u \text{ is periodic.} \end{cases} \tag{3.70}$$

By choosing proper physical units, we may assume that the resonator size is of order one. We assume that the wave speeds outside and inside the resonators are comparable to each other and that condition (3.3) holds.

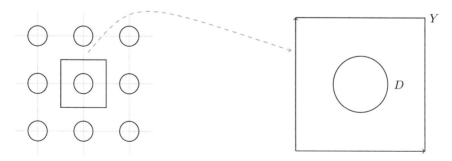

Fig. 3.5 Illustration of the square lattice crystal and quantities in Y

3.5.4   Subwavelength Band Gaps and Bloch Modes

As described in Sect. 3.5.1, the problem (3.70) has nontrivial solutions for discrete values of ω such as

$$0 \leq \omega_1^\alpha \leq \omega_2^\alpha \leq \cdots$$

and we have the following band structure of propagating frequencies for the given periodic structure:

$$[0, \max_\alpha \omega_1^\alpha] \cup [\min_\alpha \omega_2^\alpha, \max_\alpha \omega_2^\alpha] \cup [\min_\alpha \omega_3^\alpha, \max_\alpha \omega_3^\alpha] \cup \cdots .$$

A non-trivial solution to this problem and its corresponding frequency is called a Bloch eigenfunction and a Bloch eigenfrequency. The Bloch eigenfrequencies ω_i^α, $i = 1, 2, \ldots$ with positive real part, seen as functions of α, are the band functions.

We use the quasi-periodic single layer potential introduced in (3.65) to represent the solution to the scattering problem (3.70) in $Y \setminus \overline{D}$. We look for a solution u of (3.70) of the form:

$$u = \begin{cases} \mathcal{S}_D^{\alpha,k}[\psi] & \text{in } Y \setminus \overline{D}, \\ \mathcal{S}_D^{k_b}[\psi_b] & \text{in } D, \end{cases} \tag{3.71}$$

for some surface potentials $\psi, \psi_b \in L^2(\partial D)$. Using the jump relations for the single layer potentials, one can show that (3.70) is equivalent to the boundary integral equation

$$\mathcal{A}(\omega, \delta)[\Psi] = 0, \tag{3.72}$$

where

$$\mathcal{A}(\omega, \delta) = \begin{pmatrix} \mathcal{S}_D^{k_b} & -\mathcal{S}_D^{\alpha,k} \\ -\frac{1}{2}I + \mathcal{K}_D^{k_b,*} & -\delta(\frac{1}{2}I + (\mathcal{K}_D^{-\alpha,k})^*) \end{pmatrix}, \quad \Psi = \begin{pmatrix} \psi_b \\ \psi \end{pmatrix}.$$

As before, we denote by

$$\mathcal{H} = L^2(\partial D) \times L^2(\partial D) \quad \text{and} \quad \mathcal{H}_1 = H^1(\partial D) \times L^2(\partial D).$$

It is clear that $\mathcal{A}(\omega, \delta)$ is a bounded linear operator from $\mathcal{H}$ to $\mathcal{H}_1$, i.e. $\mathcal{A}(\omega, \delta) \in \mathcal{L}(\mathcal{H}, \mathcal{H}_1)$. We first look at the limiting case when $\delta = 0$. The operator $\mathcal{A}(\omega, \delta)$ is a perturbation of

$$A(\omega, 0) = \begin{pmatrix} \mathcal{S}_D^{k_b} & -\mathcal{S}_D^{\alpha,k} \\ -\frac{1}{2}I + \mathcal{K}_D^{k_b,*} & 0 \end{pmatrix}. \tag{3.73}$$

We see that ω_0 is a characteristic value of $A(\omega, 0)$ if and only if $(\omega_0 v_b^{-1})^2$ is a Neumann eigenvalue of D or $(\omega_0 v^{-1})^2$ is a Dirichlet eigenvalue of $Y \backslash \overline{D}$ with α-quasi-periodicity on ∂Y. Since zero is a Neumann eigenvalue of D, 0 is a characteristic value for the holomorphic operator-valued function $A(\omega, 0)$. By noting that there is a positive lower bound for the other Neumann eigenvalues of D and all the Dirichlet eigenvalues of $Y \backslash \overline{D}$ with α-quasi-periodicity on ∂Y, we can conclude the following result by the Gohberg-Sigal theory.

Lemma 3.5.1 *For any δ sufficiently small, there exists one and only one characteristic value $\omega_0 = \omega_0(\delta)$ in a neighborhood of the origin in the complex plane to the holomorphic operator-valued function $A(\omega, \delta)$. Moreover, $\omega_0(0) = 0$ and ω_0 depends on δ continuously.*

Asymptotic Behavior of the First Bloch Eigenfrequency ω_1^α

In this section we assume $\alpha \neq 0$. We define

$$A_0 := A(0, 0) = \begin{pmatrix} \mathcal{S}_D & -\mathcal{S}_D^{\alpha,0} \\ -\frac{1}{2}I + \mathcal{K}_D^* & 0 \end{pmatrix}, \tag{3.74}$$

and let $A_0^* : \mathcal{H}_1 \to \mathcal{H}$ be the adjoint of A_0. We choose an element $\psi_0 \in L^2(\partial D)$ such that

$$\left(-\frac{1}{2}I + \mathcal{K}_D^* \right)[\psi_0] = 0, \quad \int_{\partial D} \psi_0 = 1.$$

We recall the definition (3.27) of the capacity of the set D, Cap_D, which is equivalent to

$$\mathcal{S}_D[\psi_0] = -\frac{1}{\text{Cap}_D} \quad \text{on } \partial D. \tag{3.75}$$

Then we can easily check that $\text{Ker}(A_0)$ and $\text{Ker}(A_0^*)$ are spanned respectively by

$$\Psi_0 = \begin{pmatrix} \psi_0 \\ \tilde{\psi}_0 \end{pmatrix} \quad \text{and} \quad \Phi_0 = \begin{pmatrix} 0 \\ 1 \end{pmatrix},$$

where $\widetilde{\psi}_0 = (\mathcal{S}_D^{\alpha,0})^{-1} \mathcal{S}_D[\psi_0]$. We now perturb $\mathcal{A}_0$ by a rank-one operator $\mathcal{P}_0$ from $\mathcal{H}$ to $\mathcal{H}_1$ given by $\mathcal{P}_0[\Psi] := (\Psi, \Psi_0)\Phi_0$, and denote it by $\widetilde{\mathcal{A}}_0 = \mathcal{A}_0 + \mathcal{P}_0$. Then the followings hold:

(i) $\widetilde{\mathcal{A}}_0[\Psi_0] = \|\Psi_0\|^2 \Phi_0$, $\widetilde{\mathcal{A}}_0^*[\Phi_0] = \|\Phi_0\|^2 \Psi_0$;
(ii) The operator $\widetilde{\mathcal{A}}_0$ and its adjoint $\widetilde{\mathcal{A}}_0^*$ are invertible in $\mathcal{L}(\mathcal{H}, \mathcal{H}_1)$ and $\mathcal{L}(\mathcal{H}_1, \mathcal{H})$, respectively.

Using (3.9), (3.10), (3.66), and (3.67), we can expand $\mathcal{A}(\omega, \delta)$ as

$$\mathcal{A}(\omega, \delta) := \mathcal{A}_0 + \mathcal{B}(\omega, \delta) = \mathcal{A}_0 + \omega\mathcal{A}_{1,0} + \omega^2\mathcal{A}_{2,0} + \omega^3\mathcal{A}_{3,0} + \delta\mathcal{A}_{0,1} + \delta\omega^2\mathcal{A}_{2,1}$$
$$+ \mathcal{O}(|\omega|^4 + |\delta\omega^3|), \tag{3.76}$$

where

$$\mathcal{A}_{1,0} = \begin{pmatrix} v_b^{-1}\mathcal{S}_{D,1} & 0 \\ 0 & 0 \end{pmatrix}, \quad \mathcal{A}_{2,0} = \begin{pmatrix} v_b^{-2}\mathcal{S}_{D,2} & -v^{-2}\mathcal{S}_{D,1}^\alpha \\ v_b^{-2}\mathcal{K}_{D,2} & 0 \end{pmatrix}, \quad \mathcal{A}_{3,0} = \begin{pmatrix} v_b^{-3}\mathcal{S}_{D,3} & 0 \\ v_b^{-3}\mathcal{K}_{D,3} & 0 \end{pmatrix},$$

$$\mathcal{A}_{0,1} = \begin{pmatrix} 0 & 0 \\ 0 & -(\frac{1}{2} + (\mathcal{K}_D^{-\alpha,0})^*) \end{pmatrix}, \quad \mathcal{A}_{2,1} = \begin{pmatrix} 0 & 0 \\ 0 & -v^{-2}\mathcal{K}_{D,1}^\alpha \end{pmatrix}.$$

From the above expansion, it follows that

$$A(\omega, \delta) = -\omega^2 \frac{v_b^{-2}|D|}{\text{Cap}_D} - \omega^3 v_b^{-3} \frac{ic_1|D|}{4\pi} + c_2\delta + \omega\delta \frac{ic_1c_2v_b^{-1}\text{Cap}_D}{4\pi} \tag{3.77}$$
$$+ \mathcal{O}(|\omega|^4 + |\delta||\omega|^2 + |\delta|^2),$$

where

$$c_1 := \frac{\|\psi_0\|^2}{\|\psi_0\|^2 + \|\widetilde{\psi}_0\|^2}, \tag{3.78}$$

and

$$c_2 := \int_{\partial D} \widetilde{\psi}_0 \left(1/2 + \mathcal{K}_D^{-\alpha,0}\right)[\chi_{\partial D}] \, d\sigma. \tag{3.79}$$

We now solve $A(\omega, \delta) = 0$. It is clear that $\delta = \mathcal{O}(\omega^2)$ and thus $\omega_0(\delta) = \mathcal{O}(\sqrt{\delta})$. We write

$$\omega_0(\delta) = a_1\delta^{\frac{1}{2}} + a_2\delta + \mathcal{O}(\delta^{\frac{3}{2}}),$$

and get

$$-\frac{v_b^{-2}|D|}{\mathrm{Cap}_D}\left(a_1\delta^{\frac{1}{2}}+a_2\delta+\mathcal{O}(\delta^{\frac{3}{2}})\right)^2-\frac{ic_1v_b^{-3}|D|}{4\pi}\left(a_1\delta^{\frac{1}{2}}+a_2\delta+\mathcal{O}(\delta^{\frac{3}{2}})\right)^3$$

$$+c_2\delta+\frac{ic_1c_2v_b^{-1}\mathrm{Cap}_D}{4\pi}\left(a_1\delta^{\frac{3}{2}}+a_2\delta^2+\mathcal{O}(\delta^{\frac{5}{2}})\right)+\mathcal{O}(\delta^2)=0.$$

From the coefficients of the δ and $\delta^{\frac{3}{2}}$ terms, we obtain

$$-a_1^2\frac{v_b^{-2}|D|}{\mathrm{Cap}_D}+c_2=0$$

and

$$2a_1a_2\frac{-v_b^{-2}|D|}{\mathrm{Cap}_D}-a_1^3\frac{ic_1v_b^{-3}|D|}{4\pi}+a_1\frac{ic_1c_2v_b^{-1}\mathrm{Cap}_D}{4\pi}=0,$$

which yields

$$a_1=\pm\sqrt{\frac{c_2\mathrm{Cap}_D}{|D|}}v_b\quad\text{and}\quad a_2=0.$$

From the definition (3.68) of the α-quasi-periodic capacity, it follows that

$$c_2=\frac{\mathrm{Cap}_{D,\alpha}}{\mathrm{Cap}_D}.$$

Therefore, the following result from [7] holds.

Theorem 3.5.2 *For $\alpha\neq0$ and sufficiently small δ, we have*

$$\omega_1^\alpha=\omega_M\sqrt{\frac{\mathrm{Cap}_{D,\alpha}}{\mathrm{Cap}_D}}+\mathcal{O}(\delta^{3/2}),\tag{3.80}$$

where ω_M is defined in (3.26) by

$$\omega_M=\sqrt{\frac{\delta\mathrm{Cap}_D}{|D|}}v_b.$$

Now from (3.80), we can see that

$$\omega_{M,\alpha} := \omega_M \sqrt{\frac{\mathrm{Cap}_{D,\alpha}}{\mathrm{Cap}_D}} \to 0$$

as $\alpha \to 0$ because

$$\left((1/2)I + (\mathcal{K}_D^{-\alpha,0})^*\right)(\mathcal{S}_D^{\alpha,0})^{-1}[\chi_{\partial D}] \to 0,$$

and so $\mathrm{Cap}_{D,\alpha} \to 0$ as $\alpha \to 0$. Moreover, it is clear that $\omega_{M,\alpha}$ lies in a small neighborhood of zero.

We define $\omega_*^1 := \max_\alpha \omega_{M,\alpha}$. Then we deduce the following result regarding a subwavelength band gap opening.

Theorem 3.5.3 *For every $\varepsilon > 0$, there exists $\delta_0 > 0$ and $\widetilde{\omega} > \omega_*^1 + \varepsilon$ such that*

$$[\omega_*^1 + \varepsilon, \widetilde{\omega}] \subset [\max_\alpha \omega_1^\alpha, \min_\alpha \omega_2^\alpha] \tag{3.81}$$

for $\delta < \delta_0$.

Proof Using $\omega_1^0 = 0$ and the continuity of ω_1^α in α and δ, we get α_0 and δ_1 such that $\omega_1^\alpha < \omega_*^1$ for every $|\alpha| < \alpha_0$ and $\delta < \delta_1$. Following the derivation of (3.80), we can check that it is valid uniformly in α as far as $|\alpha| \geq \alpha_0$. Thus there exists $\delta_0 < \delta_1$ such $\omega_1^\alpha \leq \omega_*^1 + \varepsilon$ for $|\alpha| \geq \alpha_0$. We have shown that $\max_\alpha \omega_1^\alpha \leq \omega_*^1 + \varepsilon$ for sufficiently small δ. To have $\min_\alpha \omega_2^\alpha > \omega_*^1 + \varepsilon$ for small δ, it is enough to check that $\mathcal{A}(\omega, \delta)$ has no small characteristic value other than ω_1^α. For α away from 0, we can see that it is true following the proof of Theorem 3.5.2. If $\alpha = 0$, we have

$$\mathcal{A}(\omega, \delta) = \mathcal{A}(\omega, 0) + \mathcal{O}(\delta), \tag{3.82}$$

near ω_2^0 with $\delta = 0$. Since $\omega_2^0 \neq 0$, we have $\omega_2^0(\delta) > \omega_*^1 + \varepsilon$ for sufficiently small δ. Finally, using the continuity of ω_2^α in α, we obtain $\min_\alpha \omega_2^\alpha > \omega_*^1 + \varepsilon$ for small δ. This completes the proof. □

As shown in [12], the first Bloch eigenvalue ω_1^α attains its maximum ω_*^1 at $\alpha_* = (\pi, \pi, \pi)$ (i.e. the corner M of the Brillouin zone). The proof relies on the variational characterization (3.69) of the quasi-periodic capacity.

Theorem 3.5.4 *Assume that D is symmetric with respect to $\{x_j = 0\}$ for $j = 1, 2, 3$. Then both $\mathrm{Cap}_{D,\alpha}$ and ω_1^α attain their maxima at $\alpha_* = (\pi, \pi, \pi)$.*

The results of Theorems 3.5.3 and 3.5.4 are illustrated in Fig. 3.6.

Next, we consider the behavior of the first Bloch eigenfunction. In [12] a high-frequency homogenization approach for subwavelength resonators has been developed. An asymptotic expansion of the Bloch eigenfunction near the critical frequency has been computed. It is proved that the eigenfunction can be decomposed

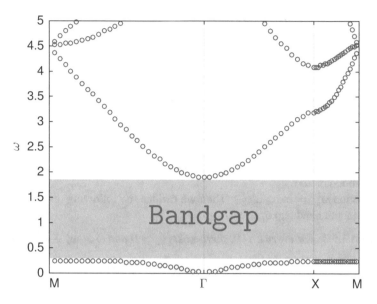

Fig. 3.6 Subwavelength band gap opening

into two parts: one is slowly varying and satisfies a homogenized equation, while the other is periodic and varying at the microscopic scale. The microscopic oscillations explain why these structures can be used to achieve super-focusing, while the exponential decay of the slowly varying part proves the band gap opening above the critical frequency.

We need the following lemma from [12].

Lemma 3.5.2 *For $\varepsilon > 0$ small enough,*

$$\mathrm{Cap}_{D,\alpha_*+\varepsilon\widetilde{\alpha}} = \mathrm{Cap}_{D,\alpha_*} + \varepsilon^2 \Lambda_D^{\widetilde{\alpha}} + \mathcal{O}(\varepsilon^4),$$

where $\Lambda_D^{\widetilde{\alpha}}$ is a negative semi-definite quadratic function of $\widetilde{\alpha}$:

$$\frac{v_b^2}{|D|}\Lambda_D^{\widetilde{\alpha}} = -\sum_{1\leq i,j\leq 3} \lambda_{ij}\widetilde{\alpha}_i\widetilde{\alpha}_j$$

with (λ_{ij}) being symmetric and positive semi-definite.

Assume that the resonators are arranged with period $r > 0$ and $\delta = \mathcal{O}(r^2)$. Then, by a scaling argument, the critical frequency $\omega_*^r = (1/r)\omega_*^1 = \mathcal{O}(1)$ as $r \to 0$.

Theorem 3.5.5 *For ω near the critical frequency ω_*^r: $(\omega_*^r)^2 - \omega^2 = \mathcal{O}(r^2)$, the following asymptotic of the first Bloch eigenfunction $u_{1,r}^{\alpha_*/r+\tilde{\alpha}}$ holds:*

$$
u_{1,r}^{\alpha_*/r+\tilde{\alpha}}(x) = \underbrace{e^{i\tilde{\alpha}\cdot x}}_{\text{macroscopic behavior}} \underbrace{S\left(\frac{x}{r}\right)}_{\text{microscopic behavior}} + \mathcal{O}(r).
$$

The macroscopic plane wave $e^{i\tilde{\alpha}\cdot x}$ satisfies:

$$
\sum_{1\leq i,j\leq 3} \lambda_{ij}\partial_i\partial_j\tilde{u}(x) + \frac{(\omega_*^r)^2 - \omega^2}{\delta}\tilde{u}(x) = 0.
$$

If we write $(\omega_*^r)^2 - \omega^2 = \beta\delta$, then

$$
\sum_{1\leq i,j\leq 3} \lambda_{ij}\tilde{\alpha}_i\tilde{\alpha}_j = \beta + \mathcal{O}(r^2).
$$

Moreover, for $\beta > 0$, the plane wave Bloch eigenfunction satisfies the homogenized equation for the crystal while the microscopic field is periodic and varies on the scale of r. If $\beta < 0$, then the Bloch eigenfunction is exponentially growing or decaying which is another way to see that a band gap opening occurs above the critical frequency.

Theorem 3.5.5 shows that the super-focusing property at subwavelength scales near the critical frequency ω_*^r holds true. Here, the mechanism is not due to effective (high-contrast below ω_*^r and negative above ω_*^r) properties of the medium. The effective medium theory described in Sect. 3.4 is no longer valid in the nondilute case.

Figure 3.7 shows a one-dimensional plot along the x_1-axis of the real part of the Bloch eigenfunction of the square lattice over many unit cells.

3.6 Topological Metamaterials

We begin this section by studying existence and consequences of a Dirac cone singularity in a honeycomb structure. Dirac singularities are intimately connected with topologically protected edge modes, and we then study such modes in an array of subwavelength resonators.

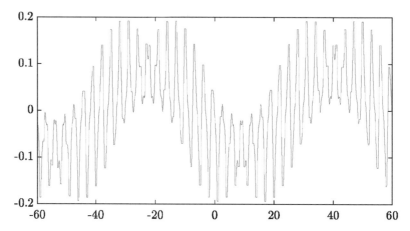

Fig. 3.7 Real part of the Bloch eigenfunction of the square lattice shown over many unit cells

3.6.1 Dirac Singularity

The classical example of a structure with a Dirac singularity is graphene, where this singularity is responsible for many peculiar electronic properties. Graphene consists of a single layer of carbon atoms in a honeycomb lattice, and in this section we study a similar structure with subwavelength resonators.

In the homogenization theory of metamaterials, the goal is to map the metamaterial to a homogeneous material with some effective parameters. It has been demonstrated in the previous section that this approach does not apply in the case of crystals at "high" frequencies, i.e., away from the centre Γ (corresponding to $\alpha = (0, 0, 0)$) of the Brillouin zone. In Theorem 3.5.5, it is shown that around the symmetry point M (corresponding to $\alpha = (\pi, \pi, \pi)$) in the Brillouin zone of a crystal with a square lattice, the Bloch eigenmodes display oscillatory behaviour on two distinct scales: small scale oscillations on the order of the size of individual resonators, while simultaneously the plane-wave envelope oscillates at a much larger scale and satisfies a homogenized equation.

In this section we prove the near-zero effective index property in a honeycomb crystal at the deep subwavelength scale. We develop a homogenization theory that captures both the macroscopic behaviour of the eigenmodes and the oscillations in the microscopic scale. The near-zero effective refractive index at the macroscale is a consequence of the existence of a Dirac dispersion cone.

We consider a two-dimensional infinite honeycomb crystal in two dimensions depicted in Fig. 3.8. Define the lattice Λ generated by the lattice vectors

$$l_1 = L\left(\frac{\sqrt{3}}{2}, \frac{1}{2}\right), \quad l_2 = L\left(\frac{\sqrt{3}}{2}, -\frac{1}{2}\right),$$

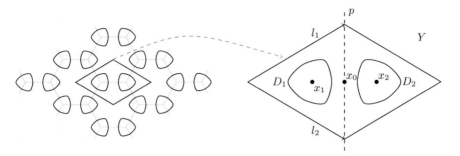

Fig. 3.8 Illustration of the honeycomb crystal and quantities in Y

where L is the lattice constant. Denote by Y a fundamental domain of the given lattice. Here, we take

$$Y := \{sl_1 + tl_2 \mid 0 \le s, t \le 1\}.$$

Define the three points x_0, x_1, and x_2 as

$$x_0 = \frac{l_1 + l_2}{2}, \quad x_1 = \frac{l_1 + l_2}{3}, \quad x_2 = \frac{2(l_1 + l_2)}{3}.$$

We will consider a general shape of the subwavelength resonators, under certain symmetry assumptions. Let R_0 be the rotation around x_0 by π, and let R_1 and R_2 be the rotations by $-\frac{2\pi}{3}$ around x_1 and x_2, respectively. These rotations can be written as

$$R_1 x = Rx + l_1, \quad R_2 x = Rx + 2l_1, \quad R_0 x = 2x_0 - x,$$

where R is the rotation by $-\frac{2\pi}{3}$ around the origin. Moreover, let R_3 be the reflection across the line $p = x_0 + \mathbb{R}e_2$, where e_2 is the second standard basis element. Assume that the unit cell contains two subwavelength resonators D_j, $j = 1, 2$, each centred at x_j such that

$$R_0 D_1 = D_2, \quad R_1 D_1 = D_1, \quad R_2 D_2 = D_2, \quad R_3 D_1 = D_2.$$

We denote the pair of subwavelength resonators by $D = D_1 \cup D_2$. The dual lattice of Λ, denoted Λ^*, is generated by α_1 and α_2 given by

$$\alpha_1 = \frac{2\pi}{L}\left(\frac{1}{\sqrt{3}}, 1\right), \quad \alpha_2 = \frac{2\pi}{L}\left(\frac{1}{\sqrt{3}}, -1\right).$$

Fig. 3.9 Illustration of the
dual lattice and the Brillouin
zone Y^*

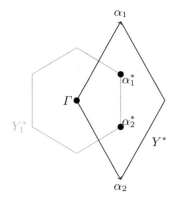

The Brillouin zone $Y^* := \mathbb{R}^2/\Lambda^*$ can be represented either as

$$Y^* \simeq \{s\alpha_1 + t\alpha_2 \mid 0 \le s, t \le 1\},$$

or as the first Brillouin zone Y_1^*, which is a hexagon illustrated in Fig. 3.9. The points

$$\alpha_1^* = \frac{2\alpha_1 + \alpha_2}{3}, \quad \alpha_2^* = \frac{\alpha_1 + 2\alpha_2}{3},$$

in the Brillouin zone are called *Dirac points*. For simplicity, we only consider the analysis around the Dirac point $\alpha_* := \alpha_1^*$, the main difference around α_2^* is summarized in Remark 3.6.1.

Wave propagation in the honeycomb lattices of subwavelength resonators is described by the following α-quasi-periodic Helmholtz problem in Y:

$$
\begin{cases}
\nabla \cdot \dfrac{1}{\rho} \nabla u + \dfrac{\omega^2}{\kappa} u = 0 & \text{in } Y \backslash \overline{D}, \\[2mm]
\nabla \cdot \dfrac{1}{\rho_b} \nabla u + \dfrac{\omega^2}{\kappa_b} u = 0 & \text{in } D, \\[2mm]
u|_+ - u|_- = 0 & \text{on } \partial D, \\[2mm]
\dfrac{1}{\rho} \dfrac{\partial u}{\partial \nu}\bigg|_+ - \dfrac{1}{\rho_b} \dfrac{\partial u}{\partial \nu}\bigg|_- = 0 & \text{on } \partial D, \\[2mm]
u(x + l) = e^{i\alpha \cdot l} u(x) & \text{for all } l \in \Lambda.
\end{cases}
\tag{3.83}
$$

Let $\psi_j^\alpha \in L^2(\partial D)$ be given by

$$\mathcal{S}_D^{\alpha,0}[\psi_j^\alpha] = \chi_{\partial D_j} \quad \text{on } \partial D, \quad j = 1, 2. \tag{3.84}$$

Define the capacitance matrix $C^\alpha = (C_{ij}^\alpha)$ by

$$C_{ij}^\alpha := -\int_{\partial D_i} \psi_j^\alpha \, d\sigma, \quad i, j = 1, 2. \tag{3.85}$$

Using the symmetry of the honeycomb structure, it can be shown that the capacitance coefficients satisfy [18]

$$c_1^\alpha := C_{11}^\alpha = C_{22}^\alpha, \quad c_2^\alpha := C_{12}^\alpha = \overline{C_{21}^\alpha},$$

and

$$\nabla_\alpha c_1^\alpha \Big|_{\alpha=\alpha^*} = 0, \quad \nabla_\alpha c_2^\alpha \Big|_{\alpha=\alpha^*} = c \begin{pmatrix} 1 \\ -i \end{pmatrix}, \tag{3.86}$$

where we denote

$$c := \frac{\partial c_2^\alpha}{\partial \alpha_1}\Big|_{\alpha=\alpha^*} \neq 0,$$

as proved in [18, Lemma 3.4].

It is shown in [18] that the first two band functions ω_1^α and ω_2^α form a conical dispersion relation near the Dirac point α_*. Such a conical dispersion is referred to as a *Dirac cone*. More specifically, the following results which hold in the subwavelength regime are proved in [18].

Theorem 3.6.1 *For small δ, the first two band functions ω_j^α, $j = 1, 2$, satisfy*

$$\omega_j^\alpha = \sqrt{\frac{\delta \lambda_j^\alpha}{|D_1|}} v_b + \mathcal{O}(\delta), \tag{3.87}$$

uniformly for α in a neighbourhood of α_*, where λ_j^α, $j = 1, 2$, are the two eigenvalues of C^α and $|D_1|$ denotes the area of one of the subwavelength resonators. Moreover, for α close to α_* and δ small enough, the first two band functions form a Dirac cone, i.e.,

$$\begin{aligned}
\omega_1^\alpha &= \omega_* - \lambda |\alpha - \alpha_*| \big[1 + \mathcal{O}(|\alpha - \alpha_*|)\big], \\
\omega_2^\alpha &= \omega_* + \lambda |\alpha - \alpha_*| \big[1 + \mathcal{O}(|\alpha - \alpha_*|)\big],
\end{aligned} \tag{3.88}$$

where ω_* and λ are independent of α and satisfy

$$\omega_* = \sqrt{\frac{\delta c_1^{\alpha_*}}{|D_1|}} v_b + \mathcal{O}(\delta) \quad \text{and} \quad \lambda = |c|\sqrt{\delta}\lambda_0 + \mathcal{O}(\delta), \quad \lambda_0 = \frac{1}{2}\sqrt{\frac{v_b^2}{|D_1| c_1^{\alpha_*}}}$$

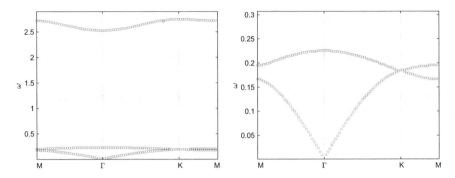

Fig. 3.10 Band gap structure upon zooming in the subwavelength region for a honeycomb lattice of subwavelength resonators

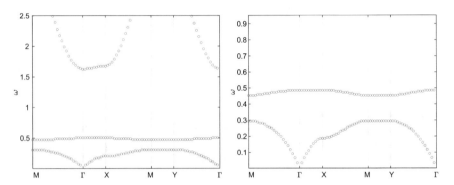

Fig. 3.11 The band gap structure upon zooming in the subwavelength region for a rectangular array of subwavelength dimers

as $\delta \to 0$. Moreover, the error term $\mathcal{O}(|\alpha - \alpha_*|)$ in (3.88) is uniform in δ.

The results in Theorem 3.6.1 are illustrated in Fig. 3.10. The figure shows the first three bands. Observe that the first two bands cross at the symmetry point K (corresponding to α_*) such that the dispersion relation is linear. Figure 3.11 shows the band gap structure in the subwavelength region for a rectangular array of subwavelength dimers. For such arrays, the two first bands cannot cross each other.

Next we investigate the asymptotic behaviour of the Bloch eigenfunctions near the Dirac points. Then we show that the envelopes of the Bloch eigenfunctions satisfy a Helmholtz equation with near-zero effective refractive index and derive a two-dimensional homogenized equation of Dirac-type for the honeycomb crystal. These results are from [19].

We consider the rescaled honeycomb crystal by replacing the lattice constant L with rL where $r > 0$ is a small positive parameter. Let ω_j^α, $j = 1, 2$, be the first two eigenvalues and u_j^α be the associated Bloch eigenfunctions for the honeycomb crystal with lattice constant L. Then, by a scaling argument, the honeycomb crystal with lattice constant rL has the first two Bloch eigenvalues

$$\omega_{\pm,r}^{\alpha/r} = \frac{1}{r}\omega_{\pm}^{\alpha},$$

and the corresponding eigenfunctions are

$$u_{\pm,r}^{\alpha/r}(x) = u_{\pm}^{\alpha}\left(\frac{x}{r}\right).$$

This shows that the Dirac cone is located at the point α_*/r. We denote the Dirac frequency by

$$\omega_*^r = \frac{1}{r}\omega_*.$$

We have the following result for the Bloch eigenfunctions $u_{j,r}^{\alpha/r}$, $j = 1, 2$, for α/r near the Dirac points α_*/r [19].

Lemma 3.6.1 *We have*

$$u_{\pm,r}^{\alpha_*/r+\tilde{\alpha}}(x) = A_{\pm}e^{i\tilde{\alpha}\cdot x}S_1\left(\frac{x}{r}\right) + B_{\pm}e^{i\tilde{\alpha}\cdot x}S_2\left(\frac{x}{r}\right) + \mathcal{O}(\delta + r),$$

where

$$S_j(x) = \mathcal{S}_D^{\alpha_*,0}[\psi_j^{\alpha_*}](x), \quad j = 1, 2.$$

The functions S_1 and S_2 describe the microscopic behaviour of the Bloch eigenfunction $u_{\pm,r}^{\alpha_*/r+\tilde{\alpha}}$ while $A_{\pm}e^{i\tilde{\alpha}\cdot x}$ and $B_{\pm}e^{i\tilde{\alpha}\cdot x}$ describe the macroscopic behaviour.

Now, we derive a homogenized equation near the Dirac frequency ω_*^r. Recall that the Dirac frequency of the unscaled honeycomb crystal satisfies $\omega_* = \mathcal{O}(\sqrt{\delta})$. As in Theorem 3.5.4, in order to make the order of ω_*^r fixed when r tends to zero, we assume that $\delta = \mu r^2$ for some fixed $\mu > 0$. Then we have

$$\omega_*^r = \frac{1}{r}\omega_* = \mathcal{O}(1) \quad \text{as } r \to 0.$$

So, in what follows, we omit the subscript r in ω_*^r, namely, $\omega_* := \omega_*^r$. Suppose the frequency ω is close to ω_*, i.e.,

$$\omega - \omega_* = \beta\sqrt{\delta} \quad \text{for some constant } \beta.$$

We need to find the Bloch eigenfunctions or $\tilde{\alpha}$ such that

$$\omega = \omega_{\pm,r}^{\alpha_*/r+\tilde{\alpha}}.$$

We have that the corresponding $\widetilde{\alpha}$ satisfies

$$\lambda_0 \begin{bmatrix} 0 & c(\widetilde{\alpha}_1 - i\widetilde{\alpha}_2) \\ \overline{c}(\widetilde{\alpha}_1 + i\widetilde{\alpha}_2) & 0 \end{bmatrix} \begin{bmatrix} A_\pm \\ B_\pm \end{bmatrix} = \beta \begin{bmatrix} A_\pm \\ B_\pm \end{bmatrix} + \mathcal{O}(s).$$

So, it is immediate to see that the macroscopic field

$$[\tilde{u}_1, \tilde{u}_2]^T := [A_\pm e^{i\widetilde{\alpha}\cdot x}, B_\pm e^{i\widetilde{\alpha}\cdot x}]^T$$

satisfies the system of Dirac equations as follows:

$$\lambda_0 \begin{bmatrix} 0 & (-ci)(\partial_1 - i\partial_2) \\ (-\overline{c}i)(\partial_1 + i\partial_2) & 0 \end{bmatrix} \begin{bmatrix} \tilde{u}_1 \\ \tilde{u}_2 \end{bmatrix} = \beta \begin{bmatrix} \tilde{u}_1 \\ \tilde{u}_2 \end{bmatrix}.$$

Here, the superscript T denotes the transpose and ∂_i is the partial derivative with respect to the ith variable. Note that the each component $\tilde{u}_j$, $j = 1, 2$, of the macroscopic field satisfies the Helmholtz equation

$$\Delta \tilde{u}_j + \frac{\beta^2}{|c|^2 \lambda_0^2} \tilde{u}_j = 0. \tag{3.89}$$

Observe, in particular, that (3.89) describes a near-zero refractive index when β is small.

The following is the main result on the homogenization theory for honeycomb lattices of subwavelength resonators [19].

Theorem 3.6.2 *For frequencies ω close to the Dirac frequency ω_*, namely, $\omega - \omega_* = \beta \sqrt{\delta}$, the following asymptotic behaviour of the Bloch eigenfunction $u_r^{\alpha_*/r + \widetilde{\alpha}}$ holds:*

$$u_r^{\alpha_*/r + \widetilde{\alpha}}(x) = A e^{i\widetilde{\alpha}\cdot x} S_1\left(\frac{x}{s}\right) + B e^{i\widetilde{\alpha}\cdot x} S_2\left(\frac{x}{s}\right) + \mathcal{O}(s),$$

where the macroscopic field

$$[\tilde{u}_1, \tilde{u}_2]^T := [A e^{i\widetilde{\alpha}\cdot x}, B e^{i\widetilde{\alpha}\cdot x}]^T$$

satisfies the two-dimensional Dirac equation

$$\lambda_0 \begin{bmatrix} 0 & (-ci)(\partial_1 - i\partial_2) \\ (-\overline{c}i)(\partial_1 + i\partial_2) & 0 \end{bmatrix} \begin{bmatrix} \tilde{u}_1 \\ \tilde{u}_2 \end{bmatrix} = \frac{\omega - \omega_*}{\sqrt{\delta}} \begin{bmatrix} \tilde{u}_1 \\ \tilde{u}_2 \end{bmatrix},$$

which can be considered as a homogenized equation for the honeycomb lattice of subwavelength resonators while the microscopic fields S_1 and S_2 vary on the scale of r.

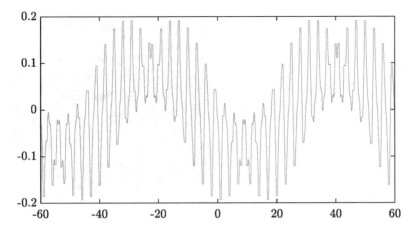

Fig. 3.12 Real part of the first Bloch eigenfunction of the honeycomb lattice shown over many unit cells

Figure 3.12 shows a one-dimensional plot along the x_1−axis of the real part of the Bloch eigenfunction of the honeycomb lattice shown over many unit cells.

Remark 3.6.1 Theorem 3.6.2 is valid around the Dirac point $\alpha_* = \alpha_1^*$. Around the other Dirac point, analogous arguments show that Theorem 3.6.2 is valid with all quantities instead defined using $\alpha_* = \alpha_2^*$ and the macroscopic field now satisfying

$$\lambda_0 \begin{bmatrix} 0 & (-ci)(\partial_1 + i\partial_2) \\ (-\bar{c}i)(\partial_1 - i\partial_2) & 0 \end{bmatrix} \begin{bmatrix} \tilde{u}_1 \\ \tilde{u}_2 \end{bmatrix} = \frac{\omega - \omega_*}{\sqrt{\delta}} \begin{bmatrix} \tilde{u}_1 \\ \tilde{u}_2 \end{bmatrix}.$$

3.6.2 Topologically Protected Edge Modes

A typical way to enable localized modes is to create a cavity inside a band gap structure. The idea is to make the frequency of the cavity mode fall within the band gap, whereby the mode will be localized to the cavity. However, localized modes created this way are highly sensitive to imperfections of the structure.

The principle that underpins the design of robust structures is that one is able to define topological invariants which capture the crystal's wave propagation properties. Then, if part of a crystalline structure is replaced with an arrangement that is associated with a different value of this invariant, not only will certain frequencies be localized to the interface but this behaviour will be stable with respect to imperfections. These eigenmodes are known as *edge modes* and we say that they are *topologically protected* to refer to their robustness.

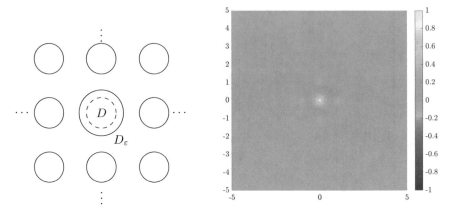

Fig. 3.13 Illustration of the defect crystal and the defect mode

Sensitivity to Geometric Imperfections

Subwavelength metamaterials can be used to achieve cavities of subwavelength dimensions. The key idea is to perturb the size of a single subwavelength resonator inside the crystal, thus creating a defect mode. Observe that if we remove one resonator inside the crystal, we cannot create a defect mode. The defect created in this fashion is actually too small to support a resonant mode. In [9], it is proved that by perturbing the radius of one resonator (see Fig. 3.13 where D_ε is the defect resonator) we create a detuned resonator with a resonant frequency that fall within the subwavelength band gap. Moreover, it is shown that the way to shift the frequency into the band gap depends on the crystal: in the dilute regime we have to decrease the defect resonator size while in the non-dilute regime we have to increase the size.

In [20], a waveguide is created by modifying the sizes of the resonators along a line in a dilute two-dimensional crystal, thereby creating a line defect. It is proved that the line defect indeed acts as a waveguide; waves of certain frequencies will be localized to, and guided along, the line defect. This is depicted in Fig. 3.14.

In wave localization due to a point defect, if the defect size is small the band structure of the defect problem will be a small perturbation of the band structure of the original problem. This way, it is possible to shift the defect band upwards, and a part of the defect band will fall into the subwavelength band gap. In [20] it is shown that for arbitrarily small defects, a part of the defect band will lie inside the band gap. Moreover, it is shown that for suitably large perturbation sizes, the entire defect band will fall into the band gap, and the size of the perturbation needed in order to achieve this can be explicitly quantified. In order to have *guided* waves along the line defect, the defect mode must not only be localized to the line, but also propagating along the line. In other words, we must exclude the case of standing waves in the line defect, *i.e.*, modes which are localized in the direction of the line. Such modes

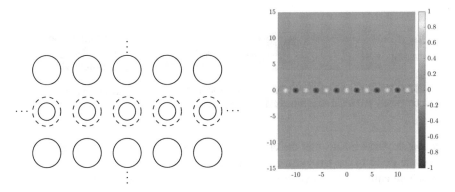

Fig. 3.14 Illustration of the line defect and the guided mode

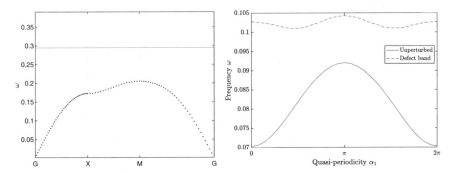

Fig. 3.15 Frequencies of the defect modes and guided waves

are associated to a point spectrum of the perturbed operator which appears as a flat band in the dispersion relation. In [20], it is shown that the defect band is nowhere flat, and hence does not correspond to bound modes in the direction of the line.

One fundamental limitation of the above designs of subwavelength cavities and waveguides is that their properties are often very sensitive to imperfections in the crystal's structure. This is due, as illustrated in Fig. 3.15, to the fact that the frequencies of the defect modes and guided waves are very close to the original band. In order to be able to feasibly manufacture wave-guiding devices, it is important that we are able to design subwavelength crystals that exhibit stability with respect to geometric errors.

Robustness Properties of One-Dimensional Chains of Subwavelength Resonators with Respect to Imperfections

In the case of one-dimensional crystals such as a chain of subwavelength resonators, the natural choice of topological invariant is the Zak phase [35]. Qualitatively, a non-

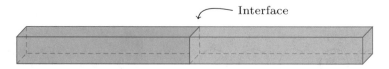

Fig. 3.16 When two crystals with different values of the topological invariant are joined together, a topologically protected edge mode exists at the interface

zero Zak phase means that the crystal has undergone *band inversion*, meaning that at some point in the Brillouin zone the monopole/dipole nature of the first/second Bloch eigenmodes has swapped. In this way, the Zak phase captures the crystal's wave propagation properties. If one takes two chains of subwavelength resonators with different Zak phases and joins half of one chain to half of the other to form a new crystal, this crystal will exhibit a topologically protected edge mode at the interface, as illustrated in Fig. 3.16.

In [15], the bulk properties of an infinitely periodic chain of subwavelength resonator dimers are studied. Using Floquet-Bloch theory, the resonant frequencies and associated eigenmodes of this crystal are derived, and further a non-trivial band gap is proved. The analogous Zak phase takes different values for different geometries and in the *dilute regime* (that is, when the distance between the resonators is an order of magnitude greater than their size) explicit expressions for its value are given. Guided by this knowledge of how the infinite (bulk) chains behave, a finite chain of resonator dimers that has a topologically protected edge mode is designed. This configuration takes inspiration from the bulk-boundary correspondence in the well-known Su-Schrieffer-Heeger (SSH) model [34] by introducing an interface, on either side of which the resonator dimers can be associated with different Zak phases thus creating a topologically protected edge mode.

In order to present the main results obtained in [15], we first briefly review the topological nature of the Bloch eigenbundle. Observe that the Brillouin zone Y^* has the topology of a circle. A natural question to ask, when considering the topological properties of a crystal, is whether properties are preserved after parallel transport around Y^*. In particular, a powerful quantity to study is the *Berry-Simon connection* A_n, defined as

$$A_n(\alpha) := i \int_D u_n^\alpha \frac{\partial}{\partial \alpha} \overline{u_n^\alpha} \, dx.$$

For any $\alpha_1, \alpha_2 \in Y^*$, the parallel transport from α_1 to α_2 is $u_n^{\alpha_1} \mapsto e^{i\theta} u_n^{\alpha_2}$, where θ is given by

$$\theta = \int_{\alpha_1}^{\alpha_2} A_n(\alpha) \, d\alpha.$$

Thus, it is enlightening to introduce the so-called *Zak phase*, φ_n^z, defined as

$$\varphi_n^z := i \int_{Y^*} \int_D u_n^\alpha \frac{\partial}{\partial \alpha} \overline{u_n^\alpha} \, dx \, d\alpha,$$

which corresponds to parallel transport around the whole of Y^*. When φ_n^z takes a value that is not a multiple of 2π, we see that the eigenmode has gained a non-zero phase after parallel transport around the circular domain Y^*. In this way, the Zak phase captures topological properties of the crystal. For crystals with inversion symmetry, the Zak phase is known to only attain the values 0 or π [35].

Next, we study a periodic arrangement of subwavelength resonator dimers. This is an analogue of the SSH model. The goal is to derive a topological invariant which characterises the crystal's wave propagation properties and indicates when it supports topologically protected edge modes.

Assume we have a one-dimensional crystal in $\mathbb{R}^3$ with repeating unit cell $Y :=$ $[-\frac{L}{2}, \frac{L}{2}] \times \mathbb{R}^2$. Each unit cell contains a dimer surrounded by some background medium. Suppose the resonators together occupy the domain $D := D_1 \cup D_2$. We need two assumptions of symmetry for the analysis that follows. The first is that each individual resonator is symmetric in the sense that there exists some $x_1 \in \mathbb{R}$ such that

$$R_1 D_1 = D_1, \quad R_2 D_2 = D_2, \tag{3.90}$$

where R_1 and R_2 are the reflections in the planes $p_1 = \{-x_1\} \times \mathbb{R}^2$ and $p_2 = \{x_1\} \times \mathbb{R}^2$, respectively. We also assume that the dimer is symmetric in the sense that

$$D = -D. \tag{3.91}$$

Denote the full crystal by $\mathcal{C}$, that is,

$$\mathcal{C} := \bigcup_{m \in \mathbb{Z}} (D + (mL, 0, 0)). \tag{3.92}$$

We denote the separation of the resonators within each unit cell, along the first coordinate axis, by $d := 2x_1$ and the separation across the boundary of the unit cell by $d' := L - d$. See Fig. 3.17.

Wave propagation inside the infinite periodic structure is modelled by the Helmholtz problem

$$\begin{cases} \Delta u + \omega^2 u = 0 & \text{in } \mathbb{R}^3 \setminus \partial\mathcal{C}, \\ u|_+ - u|_- = 0 & \text{on } \partial\mathcal{C}, \\ \delta \frac{\partial u}{\partial \nu}\Big|_+ - \frac{\partial u}{\partial \nu}\Big|_- = 0 & \text{on } \partial\mathcal{C}, \\ u(x_1, x_2, x_3) & \text{satisfies the outgoing radiation condition as } \sqrt{x_2^2 + x_3^2} \to \infty. \end{cases} \tag{3.93}$$

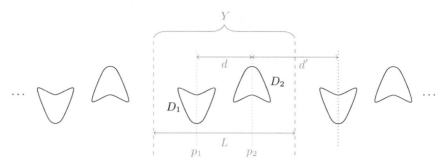

Fig. 3.17 Example of a two-dimensional cross-section of a chain of subwavelength resonators satisfying the symmetry assumptions (3.90) and (3.91). The repeating unit cell Y contains the dimer $D_1 \cup D_2$

By applying the Floquet transform, the Bloch eigenmode $u_\alpha(x) := \mathcal{U}[u](x, \alpha)$ is the solution to the Helmholtz problem

$$
\begin{cases}
\Delta u_\alpha + \omega^2 u_\alpha = 0 & \text{in } \mathbb{R}^3 \setminus \partial\mathcal{C}, \\
u_\alpha|_+ - u_\alpha|_- = 0 & \text{on } \partial\mathcal{C}, \\
\delta \dfrac{\partial u_\alpha}{\partial \nu}\bigg|_+ - \dfrac{\partial u_\alpha}{\partial \nu}\bigg|_- = 0 & \text{on } \partial\mathcal{C}, \\
e^{-i\alpha_1 x_1} u_\alpha(x_1, x_2, x_3) & \text{is periodic in } x_1, \\
u_\alpha(x_1, x_2, x_3) & \text{satisfies the } \alpha\text{-quasi-periodic outgoing radiation condition} \\
& \text{as } \sqrt{x_2^2 + x_3^2} \to \infty.
\end{cases}
\tag{3.94}
$$

We formulate the quasi-periodic resonance problem as an integral equation. Let $\mathcal{S}_D^{\alpha,\omega}$ be the single layer potential associated to the three-dimensional Green's function which is quasi-periodic in one dimension,

$$
G^{\alpha,k}(x, y) := -\sum_{m \in \mathbb{Z}} \frac{e^{ik|x - y - (Lm,0,0)|}}{4\pi |x - y - (Lm, 0, 0)|} e^{i\alpha L m}.
$$

The solution u_α of (3.94) can be represented as

$$
u_\alpha = \mathcal{S}_D^{\alpha,\omega}[\Psi^\alpha],
$$

for some density $\Psi^\alpha \in L^2(\partial D)$. Then, using the jump relations, it can be shown that (3.94) is equivalent to the boundary integral equation

$$
\mathcal{A}^\alpha(\omega, \delta)[\Psi^\alpha] = 0,
\tag{3.95}
$$

where

$$\mathcal{A}^\alpha(\omega, \delta) := -\lambda I + \left(\mathcal{K}_D^{-\alpha,\omega}\right)^*, \quad \lambda := \frac{1+\delta}{2(1-\delta)}. \tag{3.96}$$

Let V_j^α be the solution to

$$\begin{cases} \Delta V_j^\alpha = 0 & \text{in} \quad Y \setminus \overline{D}, \\ V_j^\alpha = \delta_{ij} & \text{on} \quad \partial D_i, \\ V_j^\alpha(x + (mL, 0, 0)) = e^{i\alpha m} V_j^\alpha(x) & \forall m \in \mathbb{Z}, \\ V_j^\alpha(x_1, x_2, x_3) = O\left(\frac{1}{\sqrt{x_2^2 + x_3^2}}\right) & \text{as} \quad \sqrt{x_2^2 + x_2^2} \to \infty, \text{ uniformly in } x_1, \end{cases} \tag{3.97}$$

where δ_{ij} is the Kronecker delta. Analogously to (3.85), we then define the quasi-periodic capacitance matrix $C^\alpha = (C_{ij}^\alpha)$ by

$$C_{ij}^\alpha := \int_{Y \setminus \overline{D}} \overline{\nabla V_i^\alpha} \cdot \nabla V_j^\alpha \, dx, \quad i, j = 1, 2. \tag{3.98}$$

Finding the eigenpairs of this matrix represents a leading order approximation to the differential problem (3.94). The following properties of C^α are useful.

Lemma 3.6.2 *The matrix C^α is Hermitian with constant diagonal, i.e.,*

$$C_{11}^\alpha = C_{22}^\alpha \in \mathbb{R}, \quad C_{12}^\alpha = \overline{C_{21}^\alpha} \in \mathbb{C}.$$

Since C^α is Hermitian, the following lemma follows directly.

Lemma 3.6.3 *The eigenvalues and corresponding eigenvectors of the quasi-periodic capacitance matrix are given by*

$$\lambda_1^\alpha = C_{11}^\alpha - |C_{12}^\alpha|, \quad \begin{pmatrix} a_1 \\ b_1 \end{pmatrix} = \frac{1}{\sqrt{2}} \begin{pmatrix} -e^{i\theta_\alpha} \\ 1 \end{pmatrix},$$

$$\lambda_2^\alpha = C_{11}^\alpha + |C_{12}^\alpha|, \quad \begin{pmatrix} a_2 \\ b_2 \end{pmatrix} = \frac{1}{\sqrt{2}} \begin{pmatrix} e^{i\theta_\alpha} \\ 1 \end{pmatrix},$$

where, for α such that $C_{12}^\alpha \neq 0$, $\theta_\alpha \in [0, 2\pi)$ is defined to be such that

$$e^{i\theta_\alpha} = \frac{C_{12}^\alpha}{|C_{12}^\alpha|}. \tag{3.99}$$

In the dilute regime, we are able to compute asymptotic expansions for the band structure and topological properties. In this regime, we assume that the resonators can be obtained by rescaling fixed domains B_1, B_2 as follows:

$$D_1 = \varepsilon B_1 - \left(\frac{d}{2}, 0, 0\right), \quad D_2 = \varepsilon B_2 + \left(\frac{d}{2}, 0, 0\right), \tag{3.100}$$

for some small parameter $\varepsilon > 0$.

Let Cap_B denote the capacity of $B = B_i$ for $i = 1$ or $i = 2$ (see (3.27) for the definition of the capacity). Due to symmetry, the capacitance is the same for the two choices $i = 1, 2$. It is easy to see that, by a scaling argument,

$$\text{Cap}_{\varepsilon B} = \varepsilon \text{Cap}_B. \tag{3.101}$$

Lemma 3.6.4 *We assume that the resonators are in the dilute regime specified by (3.100). We also assume that $\alpha \neq 0$ is fixed. Then we have the following asymptotics of the capacitance matrix C_{ij}^α as $\varepsilon \to 0$:*

$$C_{11}^\alpha = \varepsilon \text{Cap}_B - \frac{(\varepsilon \text{Cap}_B)^2}{4\pi} \sum_{m \neq 0} \frac{e^{im\alpha L}}{|mL|} + \mathcal{O}(\varepsilon^3), \tag{3.102}$$

$$C_{12}^\alpha = -\frac{(\varepsilon \text{Cap}_B)^2}{4\pi} \sum_{m=-\infty}^{\infty} \frac{e^{im\alpha L}}{|mL + d|} + \mathcal{O}(\varepsilon^3). \tag{3.103}$$

Taking the imaginary part of (3.103), the corresponding asymptotic formula holds uniformly in $\alpha \in Y^*$.

Define normalized extensions of V_j^α as

$$S_j^\alpha(x) := \begin{cases} \frac{1}{\sqrt{|D_1|}} \delta_{ij} & x \in D_i, \ i = 1, 2, \\ \frac{1}{\sqrt{|D_1|}} V_j^\alpha(x) & x \in Y \setminus \overline{D}, \end{cases}$$

where $|D_1|$ is the volume of one of the resonators ($|D_1| = |D_2|$ thanks to the dimer's symmetry (3.91)). The following two approximation results hold.

Theorem 3.6.3 *The characteristic values $\omega_j^\alpha = \omega_j^\alpha(\delta)$, $j = 1, 2$, of the operator $\mathcal{A}^\alpha(\omega, \delta)$, defined in (3.96), can be approximated as*

$$\omega_j^\alpha = \sqrt{\frac{\delta \lambda_j^\alpha}{|D_1|}} + \mathcal{O}(\delta),$$

where λ_j^α, $j = 1, 2$, are eigenvalues of the quasi-periodic capacitance matrix C^α.

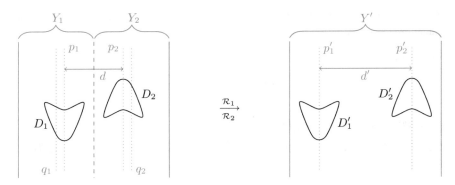

Fig. 3.18 Reflections taking D to D'

Theorem 3.6.4 *The Bloch eigenmodes* u_j^α, $j = 1, 2$, *corresponding to the resonances* ω_j^α, *can be approximated as*

$$u_j^\alpha(x) = a_j S_1^\alpha(x) + b_j S_2^\alpha(x) + \mathcal{O}(\delta),$$

where $\begin{pmatrix} a_j \\ b_j \end{pmatrix}$, $j = 1, 2$, *are the eigenvectors of the quasi-periodic capacitance matrix* C^α, *as given by Lemma 3.6.3.*

Theorems 3.6.3 and 3.6.4 show that the capacitance matrix can be considered to be a discrete approximation of the differential problem (3.94), since its eigenpairs directly determine the resonant frequencies and the Bloch eigenmodes (at leading order in δ).

We now introduce notation which, thanks to the assumed symmetry of the resonators, will allow us to prove topological properties of the chain. Divide Y into two subsets $Y = Y_1 \cup Y_2$, where $Y_1 := [-\frac{L}{2}, 0] \times \mathbb{R}^2$ and let $Y_2 := [0, \frac{L}{2}] \times \mathbb{R}^2$, as depicted in Fig. 3.18. Define q_1 and q_2 to be the central planes of Y_1 and Y_2, that is, the planes $q_1 := \{-\frac{L}{4}\} \times \mathbb{R}^2$ and $q_2 := \{\frac{L}{4}\} \times \mathbb{R}^2$. Let $\mathcal{R}_1$ and $\mathcal{R}_2$ be reflections in the respective planes. Observe that, thanks to the assumed symmetry of each resonator (3.90), the "complementary" dimer $D' = D_1' \cup D_2'$, given by swapping d and d', satisfies $D_i' = \mathcal{R}_i D_i$ for $i = 1, 2$. Define the operator T_α on the set of α-quasi-periodic functions f on Y as

$$T_\alpha f(x) := \begin{cases} e^{-i\alpha L} \overline{f(\mathcal{R}_1 x)}, & x \in Y_1, \\ \overline{f(\mathcal{R}_2 x)}, & x \in Y_2, \end{cases}$$

where the factor $e^{-i\alpha L}$ is chosen so that the image of a continuous (α-quasi-periodic) function is continuous.

We now proceed to use T_α to analyse the different topological properties of the two dimer configurations. Define the quantity $C_{12}^{\alpha\,'}$ analogously to C_{12}^α but on the dimer D', that is, to be the top-right element of the corresponding quasi-periodic capacitance matrix, defined in (3.98).

Lemma 3.6.5 *We have*

$$C_{12}^{\alpha\,'} = e^{-i\alpha L}\overline{C_{12}^{\alpha}}.$$

Consequently, if $d = d' = \frac{L}{2}$ then $C_{12}^{\pi/L} = 0$.

Lemma 3.6.6 *We assume that D is in the dilute regime specified by (3.100). Then, for ε small enough,*

(i) $\operatorname{Im} C_{12}^{\alpha} > 0$ *for* $0 < \alpha < \pi/L$ *and* $\operatorname{Im} C_{12}^{\alpha} < 0$ *for* $-\pi/L < \alpha < 0$. *In particular, $\operatorname{Im} C_{12}^{\alpha}$ is zero if and only if $\alpha \in \{0, \pi/L\}$.*

(ii) C_{12}^{α} *is zero if and only if both $d = d'$ and $\alpha = \pi/L$.*

(iii) $C_{12}^{\pi/L} < 0$ *when $d < d'$ and $C_{12}^{\pi/L} > 0$ when $d > d'$. In both cases we have $C_{12}^{0} < 0$.*

This lemma describes the crucial properties of the behaviour of the curve $\{C_{12}^{\alpha} : \alpha \in Y^*\}$ in the complex plane. The periodic nature of Y^* means that this is a closed curve. Part (i) tells us that this curve crosses the real axis in precisely two points. Taken together with (iii), we know that this curve winds around the origin in the case $d > d'$, but not in the case $d < d'$. The following band gap result is from [13].

Theorem 3.6.5 *If $d \neq d'$, the first and second bands form a band gap:*

$$\max_{\alpha \in Y^*} \omega_1^{\alpha} < \min_{\alpha \in Y^*} \omega_2^{\alpha},$$

for small enough ε and δ.

Combining the above results, we obtain the following result concerning the band inversion that takes place between the two geometric regimes $d < d'$ and $d > d'$ as illustrated in Fig. 3.19.

Proposition 3.6.1 *For ε small enough, the band structure at $\alpha = \pi/L$ is inverted between the $d < d'$ and $d > d'$ regimes. In other words, the eigenfunctions associated with the first and second bands at $\alpha = \pi/L$ are given, respectively, by*

$$u_1^{\pi/L}(x) = S_1^{\pi/L}(x) + S_2^{\pi/L}(x) + \mathcal{O}(\delta), \quad u_2^{\pi/L}(x) = S_1^{\pi/L}(x) - S_2^{\pi/L}(x) + \mathcal{O}(\delta),$$

when $d < d'$ and by

$$u_1^{\pi/L}(x) = S_1^{\pi/L}(x) - S_2^{\pi/L}(x) + \mathcal{O}(\delta), \quad u_2^{\pi/L}(x) = S_1^{\pi/L}(x) + S_2^{\pi/L}(x) + \mathcal{O}(\delta),$$

when $d > d'$.

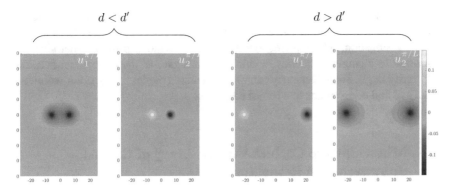

Fig. 3.19 Band inversion: the monopole/dipole natures of the 1st and 2nd eigenmodes have swapped between the $d < d'$ and $d > d'$ regimes

The eigenmode $S_1^{\pi/L}(x) + S_2^{\pi/L}(x)$ is constant and attains the same value on both resonators, while the eigenmode $S_1^{\pi/L}(x) - S_2^{\pi/L}(x)$ has values of opposite sign on the two resonators. They therefore correspond, respectively, to monopole and dipole modes, and Proposition 3.6.1 shows that the monopole/dipole nature of the first two Bloch eigenmodes are swapped between the two regimes. We will now proceed to define a topological invariant which we will use to characterise the topology of a chain and prove how its value depends on the relative sizes of d and d'. This invariant is intimately connected with the band inversion phenomenon and is non-trivial only if $d > d'$ [15].

Theorem 3.6.6 *We assume that D is in the dilute regime specified by (3.100). Then the Zak phase φ_j^z, $j = 1, 2$, defined by*

$$\varphi_j^z := \mathrm{i} \int_{Y^*} \int_D u_j^\alpha \frac{\partial}{\partial \alpha} \overline{u_j^\alpha} \, \mathrm{d}x \, \mathrm{d}\alpha,$$

satisfies

$$\varphi_j^z = \begin{cases} 0, & \text{if } d < d', \\ \pi, & \text{if } d > d', \end{cases}$$

for ε and δ small enough.

Theorem 3.6.6 shows that the Zak phase of the crystal is non-zero precisely when $d > d'$. The bulk-boundary correspondence suggests that we can create topologically protected subwavelength edge modes by joining half-space subwavelength crystals, one with $\varphi_j^z = 0$ and the other with $\varphi_j^z = \pi$.

Remark 3.6.2 A second approach to creating chains with robust subwavelength localized modes is to start with a one-dimensional array of pairs of subwavelength resonators that exhibits a subwavelength band gap. We then introduce a defect by adding a dislocation within one of the resonator pairs. As shown in [13], as a result of this dislocation, mid-gap frequencies enter the band gap from either side and converge to a single frequency, within the band gap, as the dislocation becomes arbitrarily large. Such frequency can place localized modes at any point within the band gap and corresponds to a robust edge modes.

3.7 Mimicking the Cochlea with an Array of Graded Subwavelength Resonators

In [1] an array of subwavelength resonators is used to design a to-scale artificial cochlea that mimics the first stage of human auditory processing and present a rigorous analysis of its properties. In order to replicate the spatial frequency separation of the cochlea, the array should have a size gradient, meaning each resonator is slightly larger than the previous, as depicted in Fig. 3.20. The size gradient is chosen so that the resonator array mimics the spatial frequency separation performed by the cochlea. In particular, the structure can reproduce the well-known (tonotopic) relationship between incident frequency and position of maximum excitation in the cochlea. This is a consequence of the asymmetry of the eigenmodes $u_n(x)$, see [1] and [3] for details.

Such graded arrays of subwavelength resonators can mimic the biomechanical properties of the cochlea, at the same scale. In [2], a modal time-domain expansion for the scattered pressure field due to such a structure is derived from first principles. It is proposed there that these modes should form the basis of a signal processing architecture. The properties of such an approach is investigated and it is shown that higher-order gammatone filters appear by cascading. Further, an approach for extracting meaningful global properties from the coefficients, tailored to the statistical properties of so-called natural sounds is proposed.

The subwavelength resonant frequencies of an array of $N = 22$ resonators computed by using the formulation (3.14)–(3.15) are shown in Fig. 3.21. This array measures 35 mm, has material parameters corresponding to air-filled resonators surrounded by water and has subwavelength resonant frequencies within the range

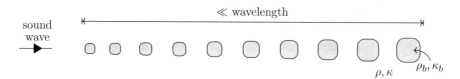

Fig. 3.20 A graded array of subwavelength resonators mimics the biomechanical properties of the cochlea in response to a sound wave

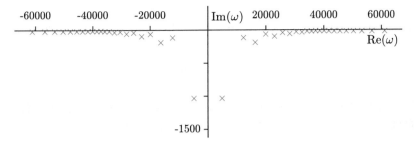

Fig. 3.21 The resonant frequencies $\{\omega_n : n = 1, \ldots, N\} \subset \mathbb{C}$ lie in the right-hand complex plane, shown for an array of $N = 22$ subwavelength resonators. The Helmholtz problem also has singularities in the left-hand plane, which are symmetric in the imaginary axis. The imaginary parts are all negative, due to energy losses

500 Hz–10 kHz. Thus, this structure has similar dimensions to the human cochlea, is made from realistic materials and experiences subwavelength resonance in response to frequencies that are audible to humans.

This analysis is useful not only for designing cochlea-like devices, but is also used in [2] as the basis for a machine hearing procedure which mimics neural processing in the auditory system. Consider the scattering of a signal, $s : [0, T] \to \mathbb{R}$, whose frequency support is wider than a single frequency and whose Fourier transform exists. Consider the Fourier transform of the incoming pressure wave, given for $\omega \in \mathbb{C}$ and $x \in \mathbb{R}^3$ by

$$
u^{in}(x, \omega) = \int_{-\infty}^{\infty} s(x_1/v - t)e^{i\omega t} \, dt
$$

$$
= e^{i\omega x_1/v}\hat{s}(\omega) = \hat{s}(\omega) + \mathcal{O}(\omega),
$$

where $\hat{s}(\omega) := \int_{-\infty}^{\infty} s(-u)e^{i\omega u} \, du$. In Theorem 3.2.1, we defined resonant frequencies as having positive real parts. However, the scattering problem (3.4) is known to be symmetric in the sense that if it has a pole at $\omega \in \mathbb{C}$ then it has a pole with the same multiplicity at $-\overline{\omega}$ [21]. As depicted in Fig. 3.21, this means the resonant spectrum is symmetric in the imaginary axis.

Suppose that the scattered acoustic pressure field u in response to the Fourier transformed signal $\hat{s}$ can, for $x \in \partial D$, be decomposed as

$$
u(x, \omega) = \sum_{n=1}^{N} \frac{-\hat{s}(\omega)v_n \operatorname{Re}(\omega_n^+)^2}{(\omega - \omega_n)(\omega + \overline{\omega_n})} u_n(x) + r(x, \omega), \tag{3.104}
$$

for some remainder r. We are interested in signals whose energy is mostly concentrated within the subwavelength regime. In particular, we want that

$$\sup_{x \in \mathbb{R}^3} \int_{-\infty}^{\infty} |r(x, \omega)| \, d\omega = \mathcal{O}(\delta). \tag{3.105}$$

Then, under the assumptions (3.104) and (3.105), we can apply the inverse Fourier transform [2] to find that the scattered pressure field satisfies, for $x \in \partial D$ and $t \in \mathbb{R}$,

$$p(x, t) = \sum_{n=1}^{N} a_n[s](t) u_n(x) + \mathcal{O}(\delta), \tag{3.106}$$

where the coefficients are given by the convolutions $a_n[s](t) = (s * h_n)(t)$ with the kernels

$$h_n(t) = \begin{cases} 0, & t < 0, \\ c_n e^{\mathrm{Im}(\omega_n)t} \sin(\mathrm{Re}(\omega_n)t), & t \geq 0, \end{cases} \tag{3.107}$$

for $c_n = v_n \, \mathrm{Re}(\omega_n)$.

Remark 3.7.1 The assumption (3.105) is a little difficult to interpret physically. For the purposes of informing signal processing approaches, however, it is sufficient.

On the one hand, note that the fact that $h_n(t) = 0$ for $t < 0$ ensures the causality of the modal expansion in (3.106). Moreover, as shown in Fig. 3.22, h_n is a windowed oscillatory mode that acts as a band-pass filter centred at $\mathrm{Re}(\omega_n)$. On the other hand, the asymmetry of the spatial eigenmodes $u_n(x)$ means that the decomposition from (3.106) replicates the cochlea's travelling wave behaviour. That is, in response to an impulse the position of maximum amplitude moves slowly from left to right in the array, see [1] for details. In [2], a signal processing architecture

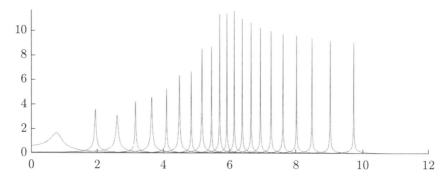

Fig. 3.22 The frequency support of the band-pass filters h_n created by an array of 22 subwavelength resonators

is developed, based on the convolutional structure of (3.106). This further mimics the action of biological auditory processing by extracting global properties of behaviourally significant sounds, to which human hearing is known to be adapted.

Finally, it is worth mentioning that biological hearing is an inherently nonlinear process. In [3] nonlinear amplification is introduced to the model in order to replicate the behaviour of the cochlear amplifier. This active structure takes the form of a fluid-coupled array of Hopf resonators. Clarifying the details of the nonlinearities that underpin cochlear function remains the largest open question in understanding biological hearing. One of the motivations for developing devices such as the one analysed here is that it will allow for the investigation of these mechanisms, which is particularly difficult to do on biological cochleas.

3.8   Concluding Remarks

In this review, recent mathematical results on focusing, trapping, and guiding waves at subwavelength scales have been described in the Hermitian case. Systems of subwavelength resonators that exhibit topologically protected edge modes or that can mimic the biomechanical properties of the cochlea have been designed. A variety of mathematical tools for solving wave propagation problems at subwavelength scales have been introduced.

When sources of energy gain and loss are introduced to a wave-scattering system, the underlying mathematical formulation will be non-Hermitian. This paves the way for new ways to control waves at subwavelength scales [22, 25, 32]. In [14, 16], the existence of asymptotic exceptional points, where eigenvalues coincide and eigenmodes are linearly dependent at leading order, in a parity–time-symmetric pair of subwavelength resonators is proved. Systems exhibiting exceptional points can be used for sensitivity enhancement. Moreover, a structure which exhibits asymptotic unidirectional reflectionless transmission at certain frequencies is designed. In [4], the phenomenon of topologically protected edge states in systems of subwavelength resonators with gain and loss is studied. It is demonstrated that localized edge modes appear in a periodic structure of subwavelength resonators with a defect in the gain/loss distribution, and the corresponding frequencies and decay lengths are explicitly computed. Similarly to the Hermitian case, these edge modes can be attributed to the winding of the eigenmodes. In the non-Hermitian case the topological invariants fail to be quantized, but can nevertheless predict the existence of localized edge modes.

The codes used for the numerical illustrations of the results described in this review can be downloaded at https://people.math.ethz.ch/~hammari/SWR.zip.

References

1. H. Ammari, B. Davies, A fully coupled subwavelength resonance approach to filtering auditory signals. Proc. R. Soc. A **475**(2228), 20190049 (2019)
2. H. Ammari, B. Davies, A biomimetic basis for auditory processing and the perception of natural sounds (2020). arXiv: 2005.12794
3. H. Ammari, B. Davies, Mimicking the active cochlea with a fluid-coupled array of subwavelength Hopf resonators. Proc. R. Soc. A **476**(2234), 20190870 (2020)
4. H. Ammari, E.O. Hiltunen, Edge modes in active systems of subwavelength resonators (2020). arXiv: 2006.05719
5. H. Ammari, H. Zhang, A mathematical theory of super-resolution by using a system of subwavelength Helmholtz resonators. Commun. Math. Phys. **337**(1), 379–428 (2015)
6. H. Ammari, H. Zhang, Effective medium theory for acoustic waves in bubbly fluids near Minnaert resonant frequency. SIAM J. Math. Anal. **49**(4), 3252–3276 (2017)
7. H. Ammari, B. Fitzpatrick, H. Lee, S. Yu, H. Zhang, Subwavelength phononic bandgap opening in bubbly media. J. Differ. Equ. **263**(9), 5610–5629 (2017)
8. H. Ammari, B. Fitzpatrick, D. Gontier, H. Lee, H. Zhang, Minnaert resonances for acoustic waves in bubbly media. Ann. I. H. Poincaré–A. N. **35**(7), 1975–1998 (2018)
9. H. Ammari, B. Fitzpatrick, E.O. Hiltunen, S. Yu, Subwavelength localized modes for acoustic waves in bubbly crystals with a defect. SIAM J. Appl. Math. **78**(6), 3316–3335 (2018)
10. H. Ammari, B. Fitzpatrick, H. Kang, M. Ruiz, S. Yu, H. Zhang, *Mathematical and Computational Methods in Photonics and Phononics*, vol. 235. Mathematical Surveys and Monographs (American Mathematical Society, Providence, 2018)
11. H. Ammari, B. Fitzpatrick, H. Lee, S. Yu, H. Zhang, Double-negative acoustic metamaterials. Quart. Appl. Math. **77**(4), 767–791 (2019)
12. H. Ammari, H. Lee, H. Zhang, Bloch waves in bubbly crystal near the first band gap: a high-frequency homogenization approach. SIAM J. Math. Anal. **51**(1), 45–59 (2019)
13. H. Ammari, B. Davies, E.O. Hiltunen, Robust edge modes in dislocated systems of subwavelength resonators. J. London Math. Soc. (2022). https://doi.org/10.1112/jlms.12619
14. H. Ammari, B. Davies, E.O. Hiltunen, H. Lee, S. Yu, High-order exceptional points and enhanced sensing in subwavelength resonator arrays. Stud. Appl. Math. **146**(2), 440–462 (2021)
15. H. Ammari, B. Davies, E.O. Hiltunen, S. Yu, Topologically protected edge modes in one-dimensional chains of subwavelength resonators. J. Math. Pures Appl. **144**, 17–49 (2020)
16. H. Ammari, B. Davies, H. Lee, E.O. Hiltunen, S. Yu, Exceptional points in parity–time-symmetric subwavelength metamaterials (2020). arXiv preprint arXiv:2003.07796
17. H. Ammari, B. Davies, S. Yu, Close-to-touching acoustic subwavelength resonators: eigenfrequency separation and gradient blow-up. Multiscale Model. Simul. **18**(3), 1299–1317 (2020)
18. H. Ammari, B. Fitzpatrick, H. Lee, E.O. Hiltunen, S. Yu, Honeycomb-lattice Minnaert bubbles. SIAM J. Math. Anal. **52**(6), 5441–5466 (2020)
19. H. Ammari, E.O. Hiltunen, S. Yu, A high-frequency homogenization approach near the Dirac points in bubbly honeycomb crystals. Arch. Rational Mech. Anal. **238**(3), 1559–1583 (2020)
20. H. Ammari, E.O. Hiltunen, S. Yu, Subwavelength guided modes for acoustic waves in bubbly crystals with a line defect. J. Eur. Math. Soc. **24**(7), 2279–2313 (2022)
21. S. Dyatlov, M. Zworski, *Mathematical Theory of Scattering Resonances*, vol. 200. Graduate Studies in Mathematics (American Mathematical Society, Providence, 2019)
22. R. El-Ganainy, K. Makris, M. Khajavikhan, Z. Musslimani, S. Rotter, D. Christodoulides, Non-Hermitian physics and PT symmetry. Nat. Phys. **14**, 11–19 (2018)
23. I. Gohberg, J. Leiterer, *Holomorphic Operator Functions of One Variable and Applications: Methods from Complex Analysis in Several Variables*, vol. 192. Operator Theory Advances and Applications (Birkhäuser, Basel, 2009)
24. I. Gohberg, E. Sigal, Operator extension of the logarithmic residue theorem and rouché's theorem. Math. USSR-Sb. **13**, 603–625 (1971)

25. Z. Gong, Y. Ashida, K. Kawabata, K. Takasan, S. Higashikawa, M. Ueda, Topological phases of non-Hermitian systems. Phys. Rev. X **8**, 031079 (2018)
26. N. Kaina, F. Lemoult, M. Fink, G. Lerosey, Negative refractive index and acoustic superlens from multiple scattering in single negative metamaterials. Nature **525**(7567), 77–81 (2015)
27. H. Kang, S. Yu, Quantitative characterization of stress concentration in the presence of closely spaced hard inclusions in two-dimensional linear elasticity. Arch. Rational Mech. Anal. **232**, 121–196 (2019)
28. P. Kuchment, An overview of periodic elliptic operators. B. Am. Math. Soc. **53**(3), 343–414 (2016)
29. M. Lanoy, R. Pierrat, F. Lemoult, M. Fink, V. Leroy, A. Tourin, Subwavelength focusing in bubbly media using broadband time reversal. Phys. Rev. B **91**, 224202 (2015)
30. J. Lekner, Near approach of two conducting spheres: Enhancement of external electric field. J. Electrostat. **69**(6), 559–563 (2011)
31. Z. Liu, X. Zhang, Y. Mao, Y. Zhu, Z. Yang, C. Chan, P. Sheng, Locally resonant sonic materials. Science **289**(5485), 1734–1736 (2000)
32. M.-A. Miri, A. Alù, Exceptional points in optics and photonics. Science **363**(6422), eaar7709 (2019)
33. J. Pendry, A chiral route to negative refraction. Science **306**, 1353–1355 (2004)
34. W.P. Su, J.R. Schrieffer, A.J. Heeger, Solitons in polyacetylene. Phys. Rev. Lett. **42**, 1698–1701 (1979)
35. J. Zak, Berry's phase for energy bands in solids. Phys. Rev. Lett. **62**, 2747–2750 (1989)

Chapter 4
Variational Methods with Application to One-Dimensional Boundary Value Problems and Numerical Evaluations

Marco Degiovanni

Abstract A Lagrangian system is considered, under one-sided growth conditions on the Lagrangian function. The existence of a critical point of mountain pass type is proved. An application to light rays is shown, with some numerical evaluations.

4.1 Introduction

After the intuitions of Gauss, Lord Kelvin, Dirichlet and Riemann in the nineteenth century and the first rigorous results, about the turn of the century, by Hilbert and Lebesgue, the *direct methods of the calculus of variations* took their modern form in the twentieth century with the work of Tonelli. The first purpose was to prove that a functional of the form

$$f(u) = \int_{s_0}^{s_1} L(s, u(s), u'(s)) \, ds$$

admits a global minimum u_{gm} when u varies in a suitable set of functions. If we suppose that

$$L : [s_0, s_1] \times \mathbb{R}^N \times \mathbb{R}^N \to \mathbb{R}$$

is continuous and that $\{\xi \mapsto L(s, x, \xi)\}$ is convex, then the usual growth condition to obtain such a result is *one-sided*, typically a suitable lower estimate of L. We refer

---

The author is a member of the Gruppo Nazionale per l'Analisi Matematica, la Probabilità e le loro Applicazioni (GNAMPA) of the Istituto Nazionale di Alta Matematica (INdAM).

M. Degiovanni (✉)
Dipartimento di Matematica e Fisica, Università Cattolica del Sacro Cuore, Brescia, Italy
e-mail: marco.degiovanni@unicatt.it

M. Chiappini, V. Vespri (eds.), *Applied Mathematical Problems in Geophysics*,
C.I.M.E. Foundation Subseries 2308, https://doi.org/10.1007/978-3-031-05321-4_4

the reader to [5] for an exhaustive presentation of the problems and the techniques involved in the study of global minima of the functional f.

A related question, when L is, say, of class C^1, is to prove that each minimum u of f satisfies the associated Euler-Lagrange equation

$$- \left[\nabla_\xi L(s, u, u') \right]' + \nabla_x L(s, u, u') = 0 \qquad \text{in } [s_0, s_1].$$

This is an easy task, under suitable assumptions on $\nabla_x L$ and $\nabla_\xi L$, but in general it is a hard question, also related to the regularity properties of u. We refer the reader to [10] for this aspect of the problem.

The study of critical points of f, not only local minima, also started in the twentieth century with the work of Birkhoff and was first developed by Morse, Ljusternik and Schnirelman. Concerning this topic, we refer the reader e.g. to [23, 26, 27]. However, when critical points are involved, the typical growth conditions on L are *two-sided*. This is due to the fact that, in the study of minima, f is usually assumed to be lower semicontinuous while, in the study of critical points, f is supposed to be of class C^1. For this reason, the classic study of critical points does not appear as a generalization of the study of minima.

In this exposition we will impose only one-sided growth conditions on L, exploiting some ideas from [17], and the functional f will be only lower semicontinuous. The structural assumptions will be described in Sect. 4.2. Nevertheless, we will obtain results in the line of critical point theory, taking advantage of the metric critical point theory developed independently in [13, 15] and in [20, 22], which will be recalled in Sect. 4.3. In Sect. 4.4 we will adapt to our setting some basic result from [5]. The main results of this exposition will be proved in Sect. 4.5, where we will show how the metric critical point theory can be actually applied to the setting described in Sect. 4.2.

A possible application concerns the variational approach to the study of light rays, according to Fermat's principle. In such a case the functional f is naturally coercive, but it may admit more local minima, which raises the question of the existence of further critical points, for instance of mountain pass type. For this reason, in Sect. 4.6 we will focus on a coercive case and prove Theorem 4.6.3, which is related to a result obtained in [19, 25] when the functional f is differentiable. Finally, in Sect. 4.7 we will consider a simple example, concerning the propagation of light in a nonhomogeneous medium, and we will provide some numerical evaluations.

4.2   Setting of the Problem

Let $N \geq 1$ and assume that

$$L : [s_0, s_1] \times \mathbb{R}^N \times \mathbb{R}^N \to \mathbb{R}$$

satisfies:

(L_1)   *for every $s \in [s_0, s_1]$, the function $\{(x, \xi) \mapsto L(s, x, \xi)\}$ is of class C^1 and the functions*

$$L : [s_0, s_1] \times \mathbb{R}^N \times \mathbb{R}^N \to \mathbb{R},$$

$$\nabla_x L, \ \nabla_\xi L : [s_0, s_1] \times \mathbb{R}^N \times \mathbb{R}^N \to \mathbb{R}^N$$

   are continuous;
   moreover, for every $(s, x) \in [s_0, s_1] \times \mathbb{R}^N$, the function $\{\xi \mapsto L(s, x, \xi)\}$ is strictly convex;
(L_2)   *for every $R, M > 0$, there exists $C_{R,M} \geq 0$ such that*

$$L(s, x, \xi) \geq M|\xi| - C_{R,M}$$

   for all $s \in [s_0, s_1]$ and $x, \xi \in \mathbb{R}^N$ with $|x| \leq R$;
(L_3)   *for every $R, \varepsilon > 0$, there exists $\delta_{R,\varepsilon} > 0$ such that*

$$L(s, x_0 + t(x_1 - x_0), \xi_0 + t(\xi_1 - \xi_0)) \leq L(s, x_0, \xi_0)$$
$$+ t \left[L(s, x_1, \xi_1) - L(s, x_0, \xi_0) \right]$$
$$+ \varepsilon t \left[1 + |L(s, x_0, \xi_0)| + |L(s, x_1, \xi_1)| \right]$$

   for all $t \in \left[0, \delta_{R,\varepsilon} \right]$, $s \in [s_0, s_1]$ and $x_0, x_1, \xi_0, \xi_1 \in \mathbb{R}^N$ with $|x_0| \leq R$, $|x_1| \leq R$ and $|x_1 - x_0| \leq \delta_{R,\varepsilon}$.

In Proposition 4.5.1 we will provide a sufficient condition to guarantee assumption (L_3).

Given $u_0, u_1 \in \mathbb{R}^N$, we are interested in the solutions u of the problem

$$\begin{cases} u \in C^1([s_0, s_1]; \mathbb{R}^N), \qquad \nabla_\xi L(s, u, u') \in C^1([s_0, s_1]; \mathbb{R}^N), \\ - \left[\nabla_\xi L(s, u, u') \right]' + \nabla_x L(s, u, u') = 0 \qquad \text{in } [s_0, s_1], \\ u(s_0) = u_0, \qquad u(s_1) = u_1. \end{cases} \qquad (P)$$

Problem (P) has a variational structure. Consider the Banach space $C([s_0, s_1]; \mathbb{R}^N)$ endowed with the sup norm $\|\cdot\|_\infty$. Taking into account (L_1) and (L_2), we can define a functional

$$f : C([s_0, s_1]; \mathbb{R}^N) \to]-\infty, +\infty]$$

by

$$f(u) = \begin{cases} \int_{s_0}^{s_1} L(s, u(s), u'(s)) \, ds & \text{if } u \in W^{1,1}(s_0, s_1; \mathbb{R}^N), u(s_0) = u_0 \\ & \text{and } u(s_1) = u_1, \\ +\infty & \text{otherwise.} \end{cases}$$

We will see in Sect. 4.5 how to obtain solutions u of (P) as "critical points" of f.

Remark 4.2.1 In the variational methods, also for nonsmooth functionals, each minimum is considered as a "critical point". Therefore, in order to obtain solutions of (P) by variational methods, it is reasonable to assume hypotheses that ensure that each minimum u of f is a solution of (P), at least in some weak sense.

Let us point out that assumptions (L_1) and (L_2) are not sufficient to guarantee that each minimum u of f satisfies the equation in (P) in the usual distributional sense, namely

$$\begin{cases} \nabla_\xi L(s, u, u') \in L^1_{loc}(]s_0, s_1[; \mathbb{R}^N), \qquad \nabla_x L(s, u, u') \in L^1_{loc}(]s_0, s_1[; \mathbb{R}^N), \\ \int_{s_0}^{s_1} \left[\nabla_\xi L(s, u, u') \cdot w' + \nabla_x L(s, u, u') \cdot w \right] ds = 0 \\ \hspace{6cm} \text{for all } w \in C^1_c(]s_0, s_1[; \mathbb{R}^N). \end{cases} \tag{4.1}$$

Actually, let

$$L : [-1, 1] \times \mathbb{R} \times \mathbb{R} \to \mathbb{R}$$

be defined as

$$L(s, x, \xi) = (s^4 - x^6)^2 \xi^{28} + \varepsilon \xi^2$$

and let $u_0 = -1, u_1 = 1$.

According to [3, Theorem 5.1], if $\varepsilon > 0$ is small enough, then the functional f admits minima and each minimum u satisfies

$$D_\xi L(s, u, u') \notin L^\infty_{loc}(]-1, 1[),$$

so that u is not a solution of (4.1).

It is easily seen that L satisfies assumptions (L_1) and (L_2), while assumption (L_3) is not satisfied, as Corollary 4.5.4 guarantees that each minimum u of f is a solution of (P).

4.3 Nonsmooth Critical Point Theory

In this section we recall some useful tools. We refer the reader to [6, 12, 13, 15, 20, 22] for proofs and more details.

Let X be a metric space endowed with the distance d. We denote by $B_\delta(u)$ the open ball of center u and radius δ. We will also consider the set $X \times \mathbb{R}$ endowed with the distance

$$d((u, \lambda), (v, \eta)) = \left(d(u, v)^2 + (\lambda - \eta)^2 \right)^{1/2} .$$

Let $f : X \to [-\infty, +\infty]$ be a function and let

$$\mathrm{epi}\,(f) = \{(u, \lambda) \in X \times \mathbb{R} : \ f(u) \le \lambda\} .$$

Definition 4.3.1 A point $u \in X$ is said to be a *local minimum* of f, if there exists a neighborhood U of u such that

$$f(w) \ge f(u) \qquad \text{for all } w \in U .$$

The next notion has been independently introduced in [13, 15] and in [22], while a variant has been developed in [20]. Here we follow the equivalent approach of [6].

Definition 4.3.2 For every $u \in X$ with $f(u) \in \mathbb{R}$, we denote by $|df|\,(u)$ the supremum of the σ's in $[0, +\infty[$ such that there exist $\delta > 0$ and a continuous map

$$\mathcal{H} : (B_\delta(u, f(u)) \cap \mathrm{epi}\,(f)) \times [0, \delta] \to X$$

satisfying

$$d(\mathcal{H}((w, \lambda), t), w) \le t , \qquad f(\mathcal{H}((w, \lambda), t)) \le \lambda - \sigma t ,$$

whenever $(w, \lambda) \in B_\delta(u, f(u)) \cap \mathrm{epi}\,(f)$ and $t \in [0, \delta]$.

The extended real number $|df|\,(u)$ is called the *weak slope* of f at u.

Remark 4.3.3 Let $u \in X$ be a local minimum of f, with $f(u) \in \mathbb{R}$. Then $|df|\,(u) = 0$.

Remark 4.3.4 Let X be an open subset of a normed space and let $f : X \to \mathbb{R}$ be of class C^1. Then we have $|df|\,(u) = \|f'(u)\|$ for all $u \in X$.

Remark 4.3.5 Let $u \in X$ with $f(u) \in \mathbb{R}$ and let (u_k) be a sequence in X converging to u with $(f(u_k))$ converging to $f(u)$.

Then we have

$$\liminf_k |df|(u_k) \geq |df|(u).$$

Definition 4.3.6 We say that $u \in X$ is a *(lower) critical point* of f if $f(u) \in \mathbb{R}$ and $|df|(u) = 0$. We say that $c \in \mathbb{R}$ is a *(lower) critical value* of f if there exists $u \in X$ such that $f(u) = c$ and $|df|(u) = 0$.

Definition 4.3.7 Let $c \in \mathbb{R}$. A sequence (u_k) in X is said to be a *Palais-Smale sequence at level c* ($(PS)_c$-*sequence*, for short) for f, if

$$\lim_k f(u_k) = c, \qquad \lim_k |df|(u_k) = 0.$$

We say that f satisfies the *Palais-Smale condition at level c* ($(PS)_c$, for short), if every $(PS)_c$-sequence for f admits a convergent subsequence in X.

The next concept was first introduced in [7], when f is smooth, and then analyzed in detail in [12], in the general case.

Definition 4.3.8 Let $\overline{u} \in X$ and $c \in \mathbb{R}$. A sequence (u_k) in X is said to be a *Cerami-Palais-Smale sequence at level c* ($(CPS)_c$-*sequence*, for short) for f, if

$$\lim_k f(u_k) = c, \qquad \lim_k \left[1 + d(u_k, \overline{u})\right] |df|(u_k) = 0.$$

We say that f satisfies the *Cerami-Palais-Smale condition at level c* ($(CPS)_c$, for short), if every $(CPS)_c$-sequence for f admits a convergent subsequence in X.

Since

$$\left[1 + d(u_k, \hat{u})\right] |df|(u_k) \leq \left[1 + d(\overline{u}, \hat{u})\right]\left[1 + d(u_k, \overline{u})\right] |df|(u_k),$$

it is easily seen that condition $(CPS)_c$ is independent of the choice of the point $\overline{u}$.

Of course, every $(CPS)_c$-sequence is a $(PS)_c$-sequence and so condition $(PS)_c$ implies $(CPS)_c$.

Several results of critical point theory can be extended to the case in which f is real valued and continuous, by means of such concepts. In view of our purposes, let us mention an extension of the celebrated mountain pass theorem (see [1, 19, 25, 26] when f is smooth).

Theorem 4.3.9 *Let X be a complete metric space and let $f : X \to \mathbb{R}$ be a continuous function. Let $u_{lm} \in X$ be a local minimum of f, let $v \in X$ with $v \neq u_{lm}$ and $f(v) \leq f(u_{lm})$ and set*

$$\Phi = \{\varphi \in C([0, 1]; X) : \varphi(0) = u_{lm}, \varphi(1) = v\}.$$

Assume that $\Phi \neq \emptyset$ and that f satisfies $(CPS)_c$ at the level

$$c = \inf_{\varphi \in \Phi} \sup_{0 \leq t \leq 1} f(\varphi(t)).$$

Then there exists a critical point u of f with $u \neq u_{lm}$, $u \neq v$ and $f(u) = c$.

Proof Mountain pass theorems when f is continuous were first proved in [13, 15] and in [22]. For this formulation, see [16, Theorem 2.9]. □

Example 4.3.10 Consider the function $f : \mathbb{R}^2 \rightarrow \mathbb{R}$ defined by

$$f(x, y) = 3x^4 - 4x^3 - 12x^2 + 33 + 12y^2.$$

It turns out that f has a local minimum at $u_{lm} = (-1, 0)$ with $f(u_{lm}) = 28$ and a global minimum at $u_{gm} = (2, 0)$ with $f(u_{gm}) = 1$. According to Theorem 4.3.9, there is a further critical point u, actually a "mountain pass point", which is $u = (0, 0)$ with $f(u) = 33$ (Figs. 4.1 and 4.2).

While [20, 22] were devoted to the case in which f is continuous, in [13, 15] also the general case was considered, taking advantage of the function $\mathcal{G}_f$ introduced in [14].

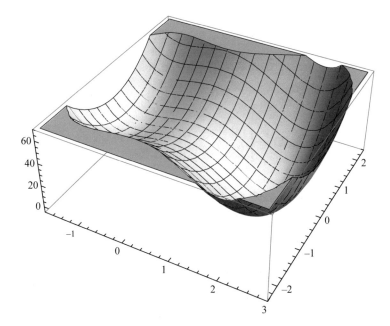

Fig. 4.1 The graph of the function f

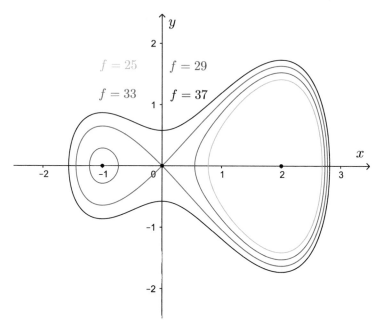

Fig. 4.2 The level lines corresponding to $f = 25$, $f = 29$, $f = 33$ and $f = 37$

For a general $f : X \rightarrow [-\infty, +\infty]$, define a function

$$\overline{\mathcal{G}_f} : X \times \mathbb{R} \rightarrow]-\infty, +\infty]$$

by

$$\overline{\mathcal{G}_f}(u, \lambda) = \begin{cases} \lambda & \text{if } (u, \lambda) \in \text{epi}(f), \\ +\infty & \text{otherwise}. \end{cases}$$

Then denote by $\mathcal{G}_f$ the restriction of $\overline{\mathcal{G}_f}$ to epi (f), which is Lipschitz continuous of constant 1, so that we have $\left|d\overline{\mathcal{G}_f}\right|(u, \lambda) = \left|d\mathcal{G}_f\right|(u, \lambda) \leq 1$ for all $(u, \lambda) \in \text{epi}(f)$.

Proposition 4.3.11 *For every $u \in X$ with $f(u) \in \mathbb{R}$, we have*

$$|df|(u) = \begin{cases} \dfrac{\left|d\mathcal{G}_f\right|(u, f(u))}{\sqrt{1 - \left|d\mathcal{G}_f\right|(u, f(u))^2}} & \text{if } \left|d\mathcal{G}_f\right|(u, f(u)) < 1, \\ +\infty & \text{if } \left|d\mathcal{G}_f\right|(u, f(u)) = 1. \end{cases}$$

Proof See [6, Proposition 2.3].   □

By means of the previous result, the study of a general f can be reduced, to some extent, to that of the continuous function $\mathcal{G}_f$. In view of the natural correspondence $u \leftrightarrow (u, f(u))$, a key point is to have a control on pairs $(u, \lambda) \in \text{epi}(f)$ with $f(u) < \lambda$.

Definition 4.3.12 Let $\overline{u} \in X$ with $f(\overline{u}) \in \mathbb{R}$ and let $c \in \mathbb{R}$. We say that f satisfies *condition (epi)$_c$*, if there exists $\varepsilon > 0$ such that

$$\inf \left\{ [1 + d((u, \lambda), (\overline{u}, f(\overline{u})))] \left| d\mathcal{G}_f \right| (u, \lambda) : \right.$$

$$\left. (u, \lambda) \in \text{epi}(f) , \ f(u) < \lambda, \ |\lambda - c| < \varepsilon \right\} > 0.$$

Again, it is easily seen that condition $(epi)_c$ is independent of the choice of $\overline{u}$.

Remark 4.3.13 If $f : X \to \mathbb{R}$ is continuous, then $\left| d\mathcal{G}_f \right| (u, \lambda) = 1$ whenever $f(u) < \lambda$.

Proof See [15, Proposition 2.3], where f is supposed to be lower semicontinious, but the assumption is not used. □

Proposition 4.3.14 *For every $c \in \mathbb{R}$, the following facts hold:*

(a) if $\mathcal{G}_f$ satisfies $(CPS)_c$, then f satisfies $(CPS)_c$;
(b) if f is lower semicontinuous and satisfies $(CPS)_c$ and $(epi)_c$, then $\mathcal{G}_f$ satisfies $(CPS)_c$.

Proof Let $\overline{u} \in X$ with $f(\overline{u}) \in \mathbb{R}$. To prove assertion (a), consider a $(CPS)_c$-sequence (u_k) for f. We have that

$$\lim_k \mathcal{G}_f(u_k, f(u_k)) = \lim_k f(u_k) = c ,$$

$$\lim_k [1 + d(u_k, \overline{u})] \, |df| (u_k) = 0 .$$

Since

$$\limsup_k \frac{1 + d((u_k, f(u_k)), (\overline{u}, f(\overline{u})))}{1 + d(u_k, \overline{u})} < +\infty ,$$

from Proposition 4.3.11 we infer that

$$\lim_k [1 + d((u_k, f(u_k)), (\overline{u}, f(\overline{u})))] \left| d\mathcal{G}_f \right| (u_k, f(u_k))$$

$$= \lim_k [1 + d((u_k, f(u_k)), (\overline{u}, f(\overline{u})))] \frac{|df| (u_k)}{\sqrt{1 + |df| (u_k)^2}} = 0 .$$

Therefore, $(u_k, f(u_k))$ is a $(CPS)_c$-sequence for $\mathcal{G}_f$. Since $\mathcal{G}_f$ satisfies $(CPS)_c$, it follows that $(u_k, f(u_k))$ admits a convergent subsequence in epi (f), so that (u_k) also admits a convergent subsequence in X.

To prove assertion (b), consider a $(CPS)_c$-sequence (u_k, λ_k) for $\mathcal{G}_f$. We have that

$$\lim_k \lambda_k = \lim_k \mathcal{G}_f(u_k, \lambda_k) = c,$$

$$\lim_k [1 + d((u_k, \lambda_k), (\overline{u}, f(\overline{u})))] \left|d\mathcal{G}_f\right|(u_k, \lambda_k) = 0.$$

From condition $(epi)_c$ we infer that $f(u_k) = \lambda_k$ eventually as $k \to \infty$, so that

$$\lim_k f(u_k) = \lim_k \lambda_k = c.$$

On the other hand, by Proposition 4.3.11 it follows that

$$[1 + d(u_k, \overline{u})] |df|(u_k) \leq [1 + d((u_k, f(u_k)), (\overline{u}, f(\overline{u})))] |df|(u_k)$$

$$= [1 + d((u_k, f(u_k)), (\overline{u}, f(\overline{u})))] \frac{\left|d\mathcal{G}_f\right|(u_k, f(u_k))}{\sqrt{1 - \left|d\mathcal{G}_f\right|(u_k, f(u_k))^2}},$$

whence

$$\lim_k [1 + d(u_k, \overline{u})] |df|(u_k) = 0,$$

namely (u_k) is a $(CPS)_c$-sequence for f. Since f satisfies $(CPS)_c$, we infer that (u_k) admits a subsequence (u_{k_j}) converging to some u in X, while (λ_k) is already convergent to c. Moreover, we have $f(u) \leq c$, as f is lower semicontinuous. Therefore (u_{k_j}, λ_{k_j}) is convergent to (u, c) in epi (f) and $\mathcal{G}_f$ satisfies $(CPS)_c$.

By the way, from Remark 4.3.5 it follows that $\left|d\mathcal{G}_f\right|(u, c) = 0$, hence that $f(u) = c$ by condition $(epi)_c$. $\qquad\square$

Theorem 4.3.15 *Let X be a complete metric space and let $f : X \to]-\infty, +\infty]$ be a lower semicontinuous function. Let $u_{lm} \in X$ be a local minimum of f with $f(u_{lm}) < +\infty$, let $v \in X$ with $v \neq u_{lm}$ and $f(v) \leq f(u_{lm})$ and set*

$$\Phi = \{\varphi \in C([0, 1]; X) : \varphi(0) = u_{lm}, \varphi(1) = v \text{ and } f \circ \varphi \text{ is bounded}\}.$$

Assume that $\Phi \neq \emptyset$ and that f satisfies $(CPS)_c$ and $(epi)_c$ at the level

$$c = \inf_{\varphi \in \Phi} \sup_{0 \leq t \leq 1} f(\varphi(t)).$$

Then there exists a critical point u of f with $u \neq u_{lm}$, $u \neq v$ and $f(u) = c$.

Proof Consider the complete metric space $\widehat{X} = \operatorname{epi}(f)$. We aim to apply Theorem 4.3.9 to the continuous function $\mathcal{G}_f : \widehat{X} \to \mathbb{R}$. We set

$$\hat{u}_{lm} = (u_{lm}, f(u_{lm})), \qquad \hat{v} = (v, f(v)),$$

$$\widehat{\Phi} = \left\{ \hat{\varphi} \in C([0,1]; \widehat{X}) : \hat{\varphi}(0) = \hat{u}_{lm}, \hat{\varphi}(1) = \hat{v} \right\},$$

so that $\hat{u}_{lm}$ is a local minimum of $\mathcal{G}_f$ and

$$\mathcal{G}_f(\hat{v}) = f(v) \leq f(u_{lm}) = \mathcal{G}_f(\hat{u}_{lm}).$$

If $\varphi \in \Phi$ and

$$b = \sup_{0 \leq t \leq 1} f(\varphi(t)),$$

we can define $\hat{\varphi} \in \widehat{\Phi}$ by

$$\hat{\varphi}(t) = \begin{cases} (u_{lm}, (1-3t)f(u_{lm}) + 3tb) & \text{if } 0 \leq t \leq 1/3, \\ (\varphi(3t-1), b) & \text{if } 1/3 \leq t \leq 2/3, \\ (v, (3-3t)b + (3t-2)f(v)) & \text{if } 2/3 \leq t \leq 1. \end{cases}$$

Since

$$\sup_{0 \leq t \leq 1} \mathcal{G}_f(\hat{\varphi}(t)) = b = \sup_{0 \leq t \leq 1} f(\varphi(t)),$$

it follows that

$$\inf_{\hat{\varphi} \in \widehat{\Phi}} \sup_{0 \leq t \leq 1} \mathcal{G}_f(\hat{\varphi}(t)) \leq c.$$

On the other hand, if $\hat{\varphi} \in \widehat{\Phi}$ and $\hat{\varphi} = (\hat{\varphi}_1, \hat{\varphi}_2)$, we have

$$f(\hat{\varphi}_1(t)) \leq \hat{\varphi}_2(t) = \mathcal{G}_f(\hat{\varphi}(t)),$$

so that $f \circ \hat{\varphi}_1$ is bounded, $\hat{\varphi}_1 \in \Phi$ and

$$\sup_{0 \leq t \leq 1} f(\hat{\varphi}_1(t)) \leq \sup_{0 \leq t \leq 1} \mathcal{G}_f(\hat{\varphi}(t)),$$

whence

$$c \leq \inf_{\hat{\varphi} \in \widehat{\Phi}} \sup_{0 \leq t \leq 1} \mathcal{G}_f(\hat{\varphi}(t)).$$

Therefore, we have

$$c = \inf_{\hat\varphi \in \hat\Phi} \sup_{0 \le t \le 1} \mathcal{G}_f(\hat\varphi(t)) .$$

From Proposition 4.3.14 we infer that $\mathcal{G}_f$ satisfies $(CPS)_c$. Then, from Theorem 4.3.9 we deduce that there exists a critical point $(u, c) \in \text{epi}(f)$ of $\mathcal{G}_f$ with $(u, c) \ne (u_{lm}, f(u_{lm}))$ and $(u, c) \ne (v, f(v))$. Again from $(epi)_c$ we infer that $f(u) = c$, so that $u \ne u_{lm}$ and $u \ne v$. By Proposition 4.3.11, u is a critical point of f. $\qquad\square$

When dealing with the weak slope $|df|(u)$, an auxiliary concept is sometimes useful. From now on in this section, we assume that X is a normed space over $\mathbb{R}$ and $f : X \to [-\infty, +\infty]$ is a function.

The next notion has been introduced in [6].

Definition 4.3.16 For every $u \in X$ with $f(u) \in \mathbb{R}$, $v \in X$ and $\varepsilon > 0$, let $f_\varepsilon^0(u; v)$ be the infimum of r's in $\mathbb{R}$ such that there exist $\delta > 0$ and a continuous map

$$\mathcal{V} : (B_\delta(u, f(u)) \cap \text{epi}(f)) \times]0, \delta] \to B_\varepsilon(v)$$

satisfying

$$f(w + t\mathcal{V}((w, \lambda), t)) \le \lambda + rt$$

whenever $(w, \lambda) \in B_\delta(u, f(u)) \cap \text{epi}(f)$ and $t \in]0, \delta]$.

Then let

$$f^0(u; v) = \sup_{\varepsilon > 0} f_\varepsilon^0(u; v) = \lim_{\varepsilon \to 0} f_\varepsilon^0(u; v) .$$

Let us recall that the function $f^0(u; \cdot) : X \to [-\infty, +\infty]$ is convex, lower semicontinuous and positively homogeneous of degree 1. Moreover $f^0(u; 0) \in \{0, -\infty\}$.

Definition 4.3.17 For every $u \in X$ with $f(u) \in \mathbb{R}$, we set

$$\partial f(u) = \left\{ \mu \in X' : \langle \mu, v \rangle \le f^0(u; v) \text{ for all } v \in X \right\} .$$

It is easily seen that $\partial f(u)$ is convex and weak* closed in X'.

Remark 4.3.18 If f is convex, then ∂f agrees with the subdifferential of convex analysis. If f is locally Lipschitz, then f^0 and ∂f agree with Clarke's notions [9], while in general $\partial_C f(u) \subseteq \partial f(u)$, where $\partial_C f(u)$ denotes Clarke's subdifferential.

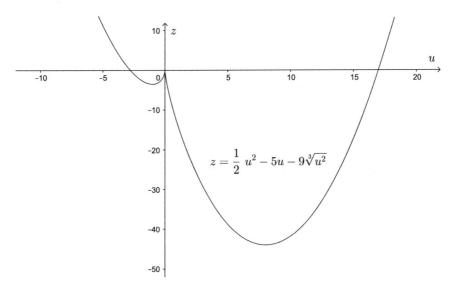

Fig. 4.3 The graph of f

Consider, for instance, the continuous function $f : \mathbb{R} \to \mathbb{R}$ defined by

$$f(u) = \frac{1}{2} u^2 - 5u - 9\sqrt[3]{u^2}.$$

It is easily seen that the assumptions of Theorem 4.3.9 are satisfied with $u_{lm} = -1$ and $v = 8$. By the way, $v = 8$ is the global minimum of f. The further critical point is $u = 0$ and in fact $|df|(0) = 0$. It follows that $0 \in \partial f(0)$, according to the next Theorem 4.3.19, while $\partial_C f(0) = \emptyset$ (Fig. 4.3).

Theorem 4.3.19 *For every $u \in X$ with $f(u) \in \mathbb{R}$, the following facts hold:*

(a) $|df|(u) < +\infty \iff \partial f(u) \neq \emptyset$;
(b) $|df|(u) < +\infty \implies |df|(u) \geq \min\{\|\mu\| : \ \mu \in \partial f(u)\}$.

Proof See [6, Theorem 4.13]. □

When f is continuous where it is finite, the Definition 4.3.16 can be simplified.

Proposition 4.3.20 *Assume there exists $D \subseteq X$ such that $f\big|_D$ is real valued and continuous, while $f = +\infty$ on $X \setminus D$.*

Then, for every $u \in D$, $v \in X$ and $\varepsilon > 0$, we have that $f_\varepsilon^0(u; v)$ is the infimum of r's in $\mathbb{R}$ such that there exist $\delta > 0$ and a continuous map

$$\mathcal{V} : (B_\delta(u) \cap D) \times]0, \delta] \to B_\varepsilon(v)$$

satisfying

$$f(w + t\mathcal{V}(w, t)) \leq f(w) + rt$$

whenever $w \in B_\delta(u) \cap D$ *and* $t \in]0, \delta]$.

Proof See [6, Proposition 4.4].   □

4.4 Compactness and Lower Semicontinuity

Throughout this section, we assume that L satisfies assumptions (L_1) and (L_2).

Theorem 4.4.1 *Let* (v_k) *be a sequence in* $W^{1,1}(s_0, s_1; \mathbb{R}^N)$ *such that*

$$\sup_k \|v_k\|_\infty < +\infty, \qquad \sup_k f(v_k) < +\infty.$$

Then there exist $u \in W^{1,1}(s_0, s_1; \mathbb{R}^N)$ *and a subsequence* (v_{k_j}) *weakly converging to* u *in* $W^{1,1}(s_0, s_1; \mathbb{R}^N)$ *with*

$$\lim_j \|v_{k_j} - u\|_\infty = 0, \qquad \liminf_j f(v_{k_j}) \geq f(u).$$

Proof Let $R > 0$ be such that

$$R \geq \sup_k \|v_k\|_\infty, \qquad R \geq \sup_k f(v_k).$$

By replacing $L(s, x, \xi)$ with $L(s, x, \xi) + C_{R,1}$, where $C_{R,1}$ is given by (L_2), we may assume that $L(s, x, \xi) \geq |\xi|$ whenever $|x| \leq R$. We infer that (v_k) is bounded in $W^{1,1}(s_0, s_1; \mathbb{R}^N)$, hence convergent, up to a subsequence we still denote by (v_k), to some u in $L^1(s_0, s_1; \mathbb{R}^N)$.

Again by (L_2), for every $\varepsilon > 0$ there exists $\widehat{C}_{R,\varepsilon} \geq 0$ such that

$$L(s, x, \xi) \geq \frac{2R}{\varepsilon} |\xi| - \widehat{C}_{R,\varepsilon} \qquad \text{whenever } |x| \leq R.$$

In particular, if c satisfies $Rc \geq \varepsilon \widehat{C}_{R,\varepsilon}$, we have

$$L(s, x, \xi) \geq \frac{R}{\varepsilon} |\xi| \qquad \text{whenever } |x| \leq R \text{ and } |\xi| \geq c.$$

It follows that

$$R \geq \int_{s_0}^{s_1} L(s, v_k, v_k') \, ds$$

$$\geq \int_{\{|v_k'| \geq c\}} L(s, v_k, v_k') \, ds$$

$$\geq \frac{R}{\varepsilon} \int_{\{|v_k'| \geq c\}} |v_k'| \, ds \,,$$

whence

$$\int_{\{|v_k'| \geq c\}} |v_k'| \, ds \leq \varepsilon \qquad \text{whenever } Rc \geq \varepsilon \widehat{C}_{R,\varepsilon} \text{ and } k \in \mathbb{N} \,.$$

According to [5, Theorem 2.12], we have that $u \in W^{1,1}(s_0, s_1; \mathbb{R}^N)$ and that (v_k) is weakly convergent to u in $W^{1,1}(s_0, s_1; \mathbb{R}^N)$, so that $u(s_0) = u_0$ and $u(s_1) = u_1$. From [5, Theorem 3.6] we also infer that

$$\liminf_k \int_{s_0}^{s_1} L(s, v_k, v_k') \, ds \geq \int_{s_0}^{s_1} L(s, u, u') \, ds \,,$$

whence

$$\liminf_k f(v_k) \geq f(u) \,.$$

Moreover, if (σ_k) is convergent to σ in $[s_0, s_1]$, then

$$|v_k(\sigma_k) - u(\sigma_k)| = \left| \int_{s_0}^{\sigma} (v_k' - u') \, ds + \int_{\sigma}^{\sigma_k} (v_k' - u') \, ds \right|$$

$$\leq \left| \int_{s_0}^{s_1} \chi_{]s_0, \sigma[} (v_k' - u') \, ds \right| + \left| \int_{\sigma}^{\sigma_k} (|v_k'| + |u'|) \, ds \right| \,.$$

Again from [5, Theorem 2.12] it follows that

$$\lim_k |v_k(\sigma_k) - u(\sigma_k)| = 0 \,,$$

whence

$$\lim_k \|v_k - u\|_\infty = 0 \,.$$

$\square$

Let us point out two obvious consequences.

Corollary 4.4.2 *The functional*

$$f : C([s_0, s_1]; \mathbb{R}^N) \to] - \infty, +\infty]$$

is lower semicontinuous.

Corollary 4.4.3 *Let $c \in \mathbb{R}$ and let (v_k) be a $(PS)_c$-sequence for f such that*

$$\sup_k \|v_k\|_\infty < +\infty .$$

Then there exist $u \in W^{1,1}(s_0, s_1; \mathbb{R}^N)$ with $f(u) \le c$ and a subsequence (v_{k_j}) such that

$$\lim_j \|v_{k_j} - u\|_\infty = 0 .$$

Theorem 4.4.4 *Let $u \in W^{1,1}(s_0, s_1; \mathbb{R}^N)$ and let (v_k) be a sequence in $W^{1,1}(s_0, s_1; \mathbb{R}^N)$ such that*

$$\lim_k \|v_k - u\|_\infty = 0 , \qquad \limsup_k f(v_k) \le f(u) < +\infty .$$

Then (v_k) is strongly convergent to u in $W^{1,1}(s_0, s_1; \mathbb{R}^N)$ and $(L(s, v_k, v_k'))$ is strongly convergent to $L(s, u, u')$ in $L^1(s_0, s_1)$.

Proof By Theorem 4.4.1 we have that (v_k) is weakly convergent to u in $W^{1,1}(s_0, s_1; \mathbb{R}^N)$. First of all, we aim to show that (v_k') is convergent to u' in measure, following an argument similar to that of [4, 28].

Let

$$R \ge \sup_k \|v_k\|_\infty .$$

Arguing as before, we may assume that $L(s, x, \xi) \ge 0$ whenever $|x| \le R$.

We claim that, for every $\varepsilon > 0$, there exists $C_\varepsilon \ge 1/\varepsilon$ such that

$$\limsup_k \int_{\{|u'| \le C_\varepsilon\}} \left[\frac{1}{2} L(s, v_k, v_k') + \frac{1}{2} L(s, v_k, u') - L\left(s, v_k, \frac{1}{2} v_k' + \frac{1}{2} u' \right) \right] ds < \varepsilon .$$

$$(4.2)$$

Actually, for every $\varepsilon > 0$, there exists $C_\varepsilon \ge 1/\varepsilon$ such that

$$\int_{s_0}^{s_1} L(s, u, u') \, ds < \int_{\{|u'| \le C_\varepsilon\}} L(s, u, u') \, ds + \varepsilon .$$

Since $L(s, v_k, v_k') \geq 0$, it follows

$$\limsup_k \int_{\{|u'| \leq C_\varepsilon\}} L(s, v_k, v_k')\, ds \leq \limsup_k \int_{s_0}^{s_1} L(s, v_k, v_k')\, ds$$

$$\leq \int_{s_0}^{s_1} L(s, u, u')\, ds$$

$$< \int_{\{|u'| \leq C_\varepsilon\}} L(s, u, u')\, ds + \varepsilon,$$

while it is easily seen that

$$\lim_k \int_{\{|u'| \leq C_\varepsilon\}} L(s, v_k, u')\, ds = \int_{\{|u'| \leq C_\varepsilon\}} L(s, u, u')\, ds.$$

Taking into account [5, Theorem 3.6], we also have

$$\liminf_k \int_{\{|u'| \leq C_\varepsilon\}} L\left(s, v_k, \frac{1}{2}v_k' + \frac{1}{2}u'\right) ds \geq \int_{\{|u'| \leq C_\varepsilon\}} L(s, u, u')\, ds$$

and (4.2) follows.

Now we claim that

$$\lim_k \left[\frac{1}{2} L(s, v_k, v_k') + \frac{1}{2} L(s, v_k, u') - L\left(s, v_k, \frac{1}{2}v_k' + \frac{1}{2}u'\right)\right] = 0 \quad \text{in measure}.$$

$$(4.3)$$

Actually, from (4.2) we infer that, for every $m \geq 1$, there exists $C_m \geq m$ such that

$$\limsup_k \int_{\{|u'| \leq C_m\}} \left[\frac{1}{2} L(s, v_k, v_k') + \frac{1}{2} L(s, v_k, u') - L\left(s, v_k, \frac{1}{2}v_k' + \frac{1}{2}u'\right)\right] ds < \frac{1}{m}.$$

Then there exists $k_m \geq m$ such that

$$\int_{\{|u'| \leq C_m\}} \left[\frac{1}{2} L(s, v_{k_m}, v_{k_m}') + \frac{1}{2} L(s, v_{k_m}, u') - L\left(s, v_{k_m}, \frac{1}{2}v_{k_m}' + \frac{1}{2}u'\right)\right] ds < \frac{1}{m},$$

whence

$$\lim_m \left\{ \chi_{\{|u'| \leq C_m\}} \left[\frac{1}{2} L(s, v_{k_m}, v_{k_m}') + \frac{1}{2} L(s, v_{k_m}, u') - L\left(s, v_{k_m}, \frac{1}{2}v_{k_m}' + \frac{1}{2}u'\right)\right]\right\} = 0$$

in $L^1(s_0, s_1)$. It follows that

$$\lim_m \left\{ \chi_{\{|u'| \le C_m\}} \left[\frac{1}{2} L(s, v_{k_m}, v'_{k_m}) + \frac{1}{2} L(s, v_{k_m}, u') - L\left(s, v_{k_m}, \frac{1}{2} v'_{k_m} + \frac{1}{2} u'\right) \right] \right\} = 0$$

in measure, whence

$$\lim_m \left[\frac{1}{2} L(s, v_{k_m}, v'_{k_m}) + \frac{1}{2} L(s, v_{k_m}, u') - L\left(s, v_{k_m}, \frac{1}{2} v'_{k_m} + \frac{1}{2} u'\right) \right] = 0$$

in measure and (4.3) follows.

Now let $\sigma > 0$. By the boundedness of (v'_k) in $L^1(s_0, s_1; \mathbb{R}^N)$, for every $\varepsilon > 0$ there exists $c > 0$ such that

$$\mathcal{L}^1 \left(\{|u'| > c\} \right) \le \frac{\varepsilon}{3}, \qquad \mathcal{L}^1 \left(\{|v'_k| > c\} \right) \le \frac{\varepsilon}{3} \qquad \text{for all } k \in \mathbb{N}. \qquad (4.4)$$

On the other hand, by the strict convexity of $\{\xi \mapsto L(s, x, \xi)\}$, there exists $\delta > 0$ such that

$$\frac{1}{2} L(s, x, \xi_0) + \frac{1}{2} L(s, x, \xi_1) - L\left(s, x, \frac{1}{2} \xi_0 + \frac{1}{2} \xi_1\right) \ge \delta$$

$$\text{whenever } |x| \le R, |\xi_0| \le c, |\xi_1| \le c \text{ and } |\xi_0 - \xi_1| \ge \sigma.$$

From (4.3) we infer that

$$\mathcal{L}^1 \left(\left\{ \frac{1}{2} L(s, v_k, v'_k) + \frac{1}{2} L(s, v_k, u') - L\left(s, v_k, \frac{1}{2} v'_k + \frac{1}{2} u'\right) \ge \delta \right\} \right) \le \frac{\varepsilon}{3}$$

eventually as $k \to \infty$, whence

$$\mathcal{L}^1 \left(\{|v'_k| \le c, \ |u'| \le c, \ |v'_k - u'| \ge \sigma\} \right) \le \frac{\varepsilon}{3}$$

eventually as $k \to \infty$. Combining this fact with (4.4), we infer that

$$\mathcal{L}^1 \left(\{|v'_k - u'| \ge \sigma\} \right) \le \varepsilon$$

eventually as $k \to \infty$, whence the convergence in measure of (v'_k) to u'.

By [18, Theorem IV.8.12] we have that (v'_k) is strongly convergent to u' in $L^1(s_0, s_1; \mathbb{R}^N)$. Since

$$|L(s, v_k, v'_k) - L(s, u, u')| \le L(s, v_k, v'_k) + L(s, u, u'),$$

from the (generalized) Lebesgue theorem we conclude that $(L(s, v_k, v'_k))$ is strongly convergent to $L(s, u, u')$ in $L^1(s_0, s_1)$. $\qquad\square$

Proposition 4.4.5 *Let $u \in W^{1,1}(s_0, s_1; \mathbb{R}^N)$ with*

$$\nabla_\xi L(s, u, u') \in L^1(s_0, s_1; \mathbb{R}^N), \quad \nabla_x L(s, u, u') \in L^1(s_0, s_1; \mathbb{R}^N).$$

Then the following facts hold:

(a) if μ belongs to the dual space of $(C([s_0, s_1]; \mathbb{R}^N), \| \ \|_\infty)$ and

$$\int_{s_0}^{s_1} \left[\nabla_\xi L(s, u, u') \cdot w' + \nabla_x L(s, u, u') \cdot w \right] ds = \langle \mu, w \rangle$$

$$\text{for all } w \in C_c^1(]s_0, s_1[; \mathbb{R}^N),$$

then we have $u \in W^{1,\infty}(s_0, s_1; \mathbb{R}^N)$, $\nabla_\xi L(s, u, u') \in BV(s_0, s_1; \mathbb{R}^N)$ and

$$\int_{s_0}^{s_1} \left[\nabla_\xi L(s, u, u') \cdot w' + \nabla_x L(s, u, u') \cdot w \right] ds = \langle \mu, w \rangle$$

$$\text{for all } w \in W_0^{1,1}(s_0, s_1; \mathbb{R}^N);$$

(b) if $z \in C([s_0, s_1]; \mathbb{R}^N)$ and

$$\int_{s_0}^{s_1} \left[\nabla_\xi L(s, u, u') \cdot w' + \nabla_x L(s, u, u') \cdot w \right] ds = \int_{s_0}^{s_1} w \cdot z \, ds$$

$$\text{for all } w \in C_c^1(]s_0, s_1[; \mathbb{R}^N),$$

then we have

$$\begin{cases} u \in C^1([s_0, s_1]; \mathbb{R}^N), & \nabla_\xi L(s, u, u') \in C^1([s_0, s_1]; \mathbb{R}^N), \\ -\left[\nabla_\xi L(s, u, u') \right]' + \nabla_x L(s, u, u') = z & \text{in } [s_0, s_1]; \end{cases}$$

if, furthermore, L is of class C^1 on $[s_0, s_1] \times \mathbb{R}^N \times \mathbb{R}^N$, then we also have

$$\begin{cases} \left[L(s, u, u') - u' \cdot \nabla_\xi L(s, u, u') \right] \in C^1([s_0, s_1]), \\ \left[L(s, u, u') - u' \cdot \nabla_\xi L(s, u, u') \right]' = D_s L(s, u, u') + u' \cdot z & \text{in } [s_0, s_1]. \end{cases}$$

Proof To prove (a), observe that of course $\nabla_\xi L(s, u, u') \in BV(s_0, s_1; \mathbb{R}^N)$ and

$$\int_{s_0}^{s_1} \left[\nabla_\xi L(s, u, u') \cdot w' + \nabla_x L(s, u, u') \cdot w \right] ds = \langle \mu, w \rangle$$

$$\text{for all } w \in W_0^{1,1}(s_0, s_1; \mathbb{R}^N).$$

In particular, there exists $K \geq \|u\|_\infty$ such that $|\nabla_\xi L(s, u, u')| \leq K/2$ a.e. in $]s_0, s_1[$. According to assumption (L_2), there exists $C_K \geq 0$ such that

$$\frac{K}{2} |u'| \geq \nabla_\xi L(s, u, u') \cdot u' \geq L(s, u, u') - L(s, u, 0) \geq K|u'| - C_K - L(s, u, 0) \,,$$

whence $u' \in L^\infty(s_0, s_1; \mathbb{R}^N)$ and assertion (a) follows.

To prove (b), observe that now $\nabla_\xi L(s, u, u') \in W^{1,\infty}(s_0, s_1; \mathbb{R}^N)$. Then there exist a continuous map $V : [s_0, s_1] \to \mathbb{R}^N$, $K > 0$ and a subset E of $]s_0, s_1[$ such that

$$\mathcal{L}^1(]s_0, s_1[\setminus E) = 0 \,, \qquad \nabla_\xi L(s, u(s), u'(s)) = V(s) \,, \ |u'(s)| \leq K \quad \text{for all } s \in E \,.$$

If (s_k), (σ_k) are to sequences in E with $(s_k - \sigma_k) \to 0$, then up to subsequences we have $s_k \to s$ and $\sigma_k \to s$ in $[s_0, s_1]$ and $u'(s_k) \to \xi$ and $u'(\sigma_k) \to \eta$ in $\mathbb{R}^N$. It follows

$$\nabla_\xi L(s, u(s), \xi) = V(s) = \nabla_\xi L(s, u(s), \eta) \,,$$

whence $\xi = \eta$, namely $(u'(s_k) - u'(\sigma_k)) \to 0$, by the strict convexity of $L(s, x, \cdot)$. Therefore the restriction of u' to E is uniformly continuous. We infer that u' agrees a.e. with a continuous map on $[s_0, s_1]$, hence that $u \in C^1([s_0, s_1]; \mathbb{R}^N)$. It follows that $\nabla_x L(s, u, u') \in C([s_0, s_1]; \mathbb{R}^N)$, $\nabla_\xi L(s, u, u') \in C^1([s_0, s_1]; \mathbb{R}^N)$ and

$$-\left[\nabla_\xi L(s, u, u') \right]' + \nabla_x L(s, u, u') = z \qquad \text{in } [s_0, s_1] \,.$$

Assume now that L is of class C^1 on $[s_0, s_1] \times \mathbb{R}^N \times \mathbb{R}^N$. We consider first the case in which $L(s, x, \xi)$ is independent of x and $z = 0$. Then we have that $\nabla_\xi L(s, u')$ is constant and, for every $v \in C^1([s_0, s_1]; \mathbb{R}^N)$ with $v(s_0) = u_0$ and $v(s_1) = u_1$, we have

$$\int_{s_0}^{s_1} L(s, v') \, ds \geq \int_{s_0}^{s_1} L(s, u') \, ds + \int_{s_0}^{s_1} \nabla_\xi L(s, u') \cdot (v' - u') \, ds = \int_{s_0}^{s_1} L(s, u') \, ds \,.$$

Let $w \in C_c^\infty(]s_0, s_1[)$ and let $\delta > 0$ be such that $\delta \|w'\|_\infty < 1$. Then there exists one and only one smooth function

$$\eta : [s_0, s_1] \times [-\delta, \delta] \to [s_0, s_1]$$

such that

$$\sigma = \eta(\sigma, t) - tw(\eta(\sigma, t)) \qquad \text{whenever } (\sigma, t) \in [s_0, s_1] \times [-\delta, \delta],$$

whence

$$\eta(\sigma, 0) = \sigma, \qquad D_t \eta(\sigma, 0) = w(\sigma) \qquad \text{whenever } \sigma \in [s_0, s_1].$$

If we set $v(s) = u(s - tw(s))$, it follows

$$\int_{s_0}^{s_1} L(s, (1 - tw'(s))u'(s - tw(s)))\, ds = \int_{s_0}^{s_1} L(s, v'(s))\, ds \geq \int_{s_0}^{s_1} L(s, u'(s))\, ds.$$

On the other hand, we have

$$\int_{s_0}^{s_1} L(s, v'(s))\, ds = \int_{s_0}^{s_1} \frac{L(s, (1 - tw'(s))u'(s - tw(s)))}{1 - tw'(s)} \cdot (1 - tw'(s))\, ds$$

$$= \int_{s_0}^{s_1} \frac{L(\eta(\sigma, t), (1 - tw'(\eta(\sigma, t)))u'(\sigma))}{1 - tw'(\eta(\sigma, t))}\, d\sigma.$$

If we set

$$I(t) = \int_{s_0}^{s_1} \frac{L(\eta(\sigma, t), (1 - tw'(\eta(\sigma, t)))u'(\sigma))}{1 - tw'(\eta(\sigma, t))}\, d\sigma,$$

then I is differentiable and $t = 0$ is a minimum point for I. On the other hand, we have

$$I'(0) = \int_{s_0}^{s_1} \Big[D_s L(\eta(\sigma, 0), u'(\sigma))\, D_t \eta(\sigma, 0)$$

$$- \nabla_\xi L(\eta(\sigma, 0), u'(\sigma)) \cdot (w'(\eta(\sigma, 0))u'(\sigma))$$

$$+ L(\eta(\sigma, 0), u'(\sigma))\, w'(\eta(\sigma, 0)) \Big]\, d\sigma$$

$$= \int_{s_0}^{s_1} \Big[D_s L(\sigma, u'(\sigma))\, w(\sigma) - \nabla_\xi L(\sigma, u'(\sigma)) \cdot (w'(\sigma)u'(\sigma)) + L(\sigma, u'(\sigma))\, w'(\sigma) \Big]\, d\sigma,$$

whence

$$\int_{s_0}^{s_1} \Big[\big(L(s, u') - u' \cdot \nabla_\xi L(s, u') \big) w' + D_s L(s, u')\, w \Big]\, ds = 0$$

$$\text{for all } w \in C_c^\infty(]s_0, s_1[).$$

Since $D_s L(s, u') \in C([s_0, s_1])$, we have that $\left(L(s, u') - u' \cdot \nabla_\xi L(s, u')\right) \in C^1([s_0, s_1])$ and

$$\left(L(s, u') - u' \cdot \nabla_\xi L(s, u')\right)' = D_s L(s, u') \qquad \text{in } [s_0, s_1].$$

In the general case, let $V \in C^1([s_0, s_1]; \mathbb{R}^N)$ be such that

$$V' = \nabla_x L(s, u, u') - z$$

and let

$$\widetilde{L}(s, \xi) = L(s, u(s), \xi) - \xi \cdot V(s).$$

Then we have

$$\nabla_\xi \widetilde{L}(s, \xi) = \nabla_\xi L(s, u(s), \xi) - V(s),$$

whence

$$\left[\nabla_\xi \widetilde{L}(s, u'(s))\right]' = \left[\nabla_\xi L(s, u(s), u'(s))\right]' - \left(\nabla_x L(s, u, u') - z\right) = 0.$$

Since

$$D_s \widetilde{L}(s, \xi) = D_s L(s, u(s), \xi) + \nabla_x L(s, u(s), \xi) \cdot u'(s) - \xi \cdot \left(\nabla_x L(s, u(s), u'(s)) - z(s)\right),$$

from the previous step we infer that

$$\begin{aligned}
\widetilde{L}(s, u') - u' \cdot \nabla_\xi \widetilde{L}(s, u') &= L(s, u, u') - u' \cdot V - u' \cdot \left(\nabla_\xi L(s, u, u') - V\right) \\
&= L(s, u, u') - u' \cdot \nabla_\xi L(s, u, u')
\end{aligned}$$

belongs to $C^1([s_0, s_1])$ and that

$$\begin{aligned}
\left[L(s, u, u') - u' \cdot \nabla_\xi L(s, u, u')\right]' &= \left[\widetilde{L}(s, u') - u' \cdot \nabla_\xi \widetilde{L}(s, u')\right]' \\
&= D_s \widetilde{L}(s, u') \\
&= D_s L(s, u, u') + \nabla_x L(s, u, u') \cdot u' \\
&\qquad - u' \cdot \left(\nabla_x L(s, u, u') - z\right) \\
&= D_s L(s, u, u') + u' \cdot z,
\end{aligned}$$

so that the proof of assertion (b) is complete. $\qquad\qquad\qquad\qquad\qquad\qquad \square$

4.5 The Variational Approach

In this section we will see how to apply the general concepts of Sect. 4.3 to the setting described in Sect. 4.2. Here the assumption (L_3) will play a crucial role. Let us first provide a sufficient condition to guarantee such a hypothesis.

Proposition 4.5.1 *Assume that L satisfies (L_1), (L_2) and that, for every $R > 0$, there exists $\widetilde{C}_R \geq 0$ such that*

$$|\nabla_x L(s, x, \xi)| \leq \widetilde{C}_R \left(1 + |L(s, x, \xi)|\right)$$

$$\text{for all } s \in [s_0, s_1] \text{ and } x, \xi \in \mathbb{R}^N \text{ with } |x| \leq R.$$

Then L satisfies assumption (L_3).

Proof Let $R > 0$. According to (L_2), we may again assume, without loss of generality, that $L(s, x, \xi) \geq 0$ whenever $|x| \leq R$.

Let now $x_0, x_1 \in \mathbb{R}^N$ with $|x_0| \leq R$ and $|x_1| \leq R$. If we set

$$\varphi(t) = L(s, x_0 + t(x_1 - x_0), \xi_0)$$

whenever $0 \leq t \leq 1$, we infer that

$$\varphi'(t) = \nabla_x L(s, x_0 + t(x_1 - x_0), \xi_0) \cdot (x_1 - x_0) \leq |\nabla_x L(s, x_0 + t(x_1 - x_0), \xi_0)| \, |x_1 - x_0|$$

$$\leq \widetilde{C}_R |x_1 - x_0|(1 + L(s, x_0 + t(x_1 - x_0), \xi_0)) = \widetilde{C}_R |x_1 - x_0|(1 + \varphi(t)),$$

whence

$$L(s, x_0 + t(x_1 - x_0), \xi_0) \leq (L(s, x_0, \xi_0) + 1) \exp(t\widetilde{C}_R |x_1 - x_0|) - 1.$$

Since

$$\exp(t\widetilde{C}_R |x_1 - x_0|) \leq 1 + t\widetilde{C}_R |x_1 - x_0| \exp(2R\widetilde{C}_R),$$

there exists $\widehat{C}_R \geq 0$ such that

$$L(s, x_0 + t(x_1 - x_0), \xi_0) \leq L(s, x_0, \xi_0) + \widehat{C}_R t|x_1 - x_0|(1 + L(s, x_0, \xi_0))$$

$$\text{whenever } 0 \leq t \leq 1.$$

Of course, we also have

$$L(s, x_0 + t(x_1 - x_0), \xi_1) = L(s, x_1 + (1 - t)(x_0 - x_1), \xi_1)$$

$$\leq L(s, x_1, \xi_1) + \widehat{C}_R(1 - t)|x_1 - x_0|(1 + L(s, x_1, \xi_1)),$$

whence

$$L(s, (1-t)x_0 + tx_1, (1-t)\xi_0 + t\xi_1)$$
$$\leq (1-t)L(s, x_0 + t(x_1 - x_0), \xi_0) + tL(s, x_0 + t(x_1 - x_0), \xi_1)$$
$$\leq (1-t)L(s, x_0, \xi_0) + tL(s, x_1, \xi_1)$$
$$+ \widehat{C}_R |x_1 - x_0| t (1-t) (2 + L(s, x_0, \xi_0) + L(s, x_1, \xi_1))$$

and the assertion easily follows.   □

From now on in this section, we assume that L satisfies assumptions (L_1)–(L_3).

Theorem 4.5.2 *For every* $u \in W^{1,1}(s_0, s_1; \mathbb{R}^N)$ *with* $f(u) < +\infty$, *there exists* $\varrho > 0$ *such that the following facts hold:*

(a) we have

$$\left[\nabla_\xi L(s, u, u') \cdot (v' - u') + \nabla_x L(s, u, u') \cdot (v - u)\right]^+ \in L^1(s_0, s_1),$$
$$\int_{s_0}^{s_1} \left[\nabla_\xi L(s, u, u') \cdot (v' - u') + \nabla_x L(s, u, u') \cdot (v - u)\right] ds \geq f^0(u; v - u),$$

for all $v \in W^{1,1}(s_0, s_1; \mathbb{R}^N)$ *with* $f(v) < +\infty$ *and* $\|v - u\|_\infty < \varrho$;

(b) if $\mu \in \partial f(u)$, *then we have*

$$\left[\nabla_\xi L(s, u, u') \cdot (v' - u') + \nabla_x L(s, u, u') \cdot (v - u)\right] \in L^1(s_0, s_1),$$
$$\int_{s_0}^{s_1} \left[\nabla_\xi L(s, u, u') \cdot (v' - u') + \nabla_x L(s, u, u') \cdot (v - u)\right] ds \geq \langle \mu, v - u \rangle,$$

for all $v \in W^{1,1}(s_0, s_1; \mathbb{R}^N)$ *with* $f(v) < +\infty$ *and* $\|v - u\|_\infty < \varrho$;

moreover, we also have $u \in W^{1,\infty}(s_0, s_1; \mathbb{R}^N)$, $\nabla_\xi L(s, u, u') \in BV(s_0, s_1; \mathbb{R}^N)$ *and*

$$\int_{s_0}^{s_1} \left[\nabla_\xi L(s, u, u') \cdot w' + \nabla_x L(s, u, u') \cdot w\right] ds = \langle \mu, w \rangle$$

for all $w \in W_0^{1,1}(s_0, s_1; \mathbb{R}^N)$.

Proof Let $R \geq 1 + \|u\|_\infty$. By (L_3) there exists $\delta_R > 0$ such that

$$L(s, x_0 + t(x_1 - x_0), \xi_0 + t(\xi_1 - \xi_0))$$
$$\leq L(s, x_0, \xi_0) + t[L(s, x_1, \xi_1) - L(s, x_0, \xi_0)]$$
$$+ t[1 + |L(s, x_0, \xi_0)| + |L(s, x_1, \xi_1)|] \qquad (4.5)$$

whenever $0 \leq t \leq \delta_R$, $|x_0| \leq R$, $|x_1| \leq R$ and $|x_1 - x_0| \leq \delta_R$, whence

$$\nabla_\xi L(s, x_0, \xi_0) \cdot (\xi_1 - \xi_0) + \nabla_x L(s, x_0, \xi_0) \cdot (x_1 - x_0)$$
$$\leq L(s, x_1, \xi_1) - L(s, x_0, \xi_0) + [1 + |L(s, x_0, \xi_0)| + |L(s, x_1, \xi_1)|] . \qquad (4.6)$$

Let $\varrho = \min\{\delta_R, 1\}$ and let $v \in W^{1,1}(s_0, s_1; \mathbb{R}^N)$ with $f(v) < +\infty$ and $\|v - u\|_\infty < \varrho$, whence $\|v\|_\infty < R$. First of all, since $L(s, u, u') \in L^1(s_0, s_1)$ and $L(s, v, v') \in L^1(s_0, s_1)$, from (4.6) we infer that

$$\left[\nabla_\xi L(s, u, u') \cdot (v' - u') + \nabla_x L(s, u, u') \cdot (v - u) \right]^+ \in L^1(s_0, s_1) .$$

Now let

$$r > \int_{s_0}^{s_1} \left[\nabla_\xi L(s, u, u') \cdot (v' - u') + \nabla_x L(s, u, u') \cdot (v - u) \right] ds .$$

We claim that there exists $\sigma > 0$ such that

$$\int_{s_0}^{s_1} \frac{L(s, z + t(v - z), z' + t(v' - z')) - L(s, z, z')}{t} ds < r$$

whenever

$$z \in W^{1,1}(s_0, s_1; \mathbb{R}^N), \qquad \|z - u\|_\infty \leq \sigma, \qquad f(z) \leq f(u) + \sigma, \qquad 0 < t \leq \sigma .$$

To prove it assume, for a contradiction, that there exist $t_k \to 0^+$ and $z_k \to u$ in $C([s_0, s_1]; \mathbb{R}^N)$ with $\limsup_k f(z_k) \leq f(u)$ satisfying

$$\int_{s_0}^{s_1} \frac{L(s, z_k + t_k(v - z_k), z_k' + t_k(v' - z_k')) - L(s, z_k, z_k')}{t_k} ds \geq r .$$

From Theorem 4.4.4 we infer that (z_k) is strongly convergent to u in $W^{1,1}(s_0, s_1; \mathbb{R}^N)$ and that $(L(s, z_k, z_k'))$ is strongly convergent to $L(s, u, u')$ in $L^1(s_0, s_1)$. Moreover, we have

$$\frac{L(s, z_k + t_k(v - z_k), z_k' + t_k(v' - z_k')) - L(s, z_k, z_k')}{t_k}$$
$$\leq L(s, v, v') - L(s, z_k, z_k') + [1 + |L(s, z_k, z_k')| + |L(s, v, v')|]$$

eventually as $k \to \infty$ by (4.5). From the (generalized) Fatou lemma, we infer that

$$r \leq \limsup_{k} \int_{s_0}^{s_1} \frac{L(s, z_k + t_k(v - z_k), z'_k + t_k(v' - z'_k)) - L(s, z_k, z'_k)}{t_k} \, ds$$

$$\leq \int_{s_0}^{s_1} \left[\nabla_\xi L(s, u, u') \cdot (v' - u') + \nabla_x L(s, u, u') \cdot (v - u) \right] ds$$

and a contradiction follows, proving the claim.

Then we have that

$$f(z + t(v - z)) \leq f(z) + rt \,,$$

whenever $z \in C([s_0, s_1]; \mathbb{R}^N)$, $\|z - u\|_\infty \leq \sigma$, $f(z) \leq f(u) + \sigma$ and $0 \leq t \leq \sigma$. Given $\varepsilon > 0$, we may also assume that $\sigma < \varepsilon$. Then, if we set

$$\mathcal{V}((z, \lambda), t) = v - z \,,$$

it follows that $\|\mathcal{V}((z, \lambda), t) - (v - u)\|_\infty < \varepsilon$ and

$$f(z + t\mathcal{V}((z, \lambda), t)) \leq f(z) + rt \leq \lambda + rt \,,$$

whenever $(z, \lambda) \in B_\sigma((u, f(u))) \cap \mathrm{epi}(f)$ and $0 \leq t \leq \sigma$. According to Definition 4.3.16, we have that

$$f_\varepsilon^0(u; v - u) \leq r$$

and assertion (a) follows by the arbitrariness of ε and r.

Assume now that $\mu \in \partial f(u)$. From assertion (a) and Definition 4.3.17 it readily follows that

$$\nabla_\xi L(s, u, u') \cdot (v' - u') + \nabla_x L(s, u, u') \cdot (v - u) \in L^1(s_0, s_1) \,,$$

$$\int_{s_0}^{s_1} \left[\nabla_\xi L(s, u, u') \cdot (v' - u') + \nabla_x L(s, u, u') \cdot (v - u) \right] ds \geq \langle \mu, v - u \rangle \,,$$

for all $v \in W^{1,1}(s_0, s_1; \mathbb{R}^N)$ with $f(v) < +\infty$ and $\|v - u\|_\infty < \varrho$.

In particular, there exists $v_0 \in C^1([s_0, s_1]; \mathbb{R}^N)$ such that $v_0(s_0) = u_0$, $v_0(s_1) = u_1$ and $\|v_0 - u\|_\infty < \varrho/2$, whence

$$\nabla_\xi L(s, u, u') \cdot (v'_0 - u') + \nabla_x L(s, u, u') \cdot (v_0 - u) \in L^1(s_0, s_1) \,.$$

Given $w \in C^1([s_0, s_1]; \mathbb{R}^N)$ with $\|w\|_\infty \leq 1$, from (4.6) we also infer that

$$\nabla_\xi L(s, u, u') \cdot ((\varrho/2)w' + v'_0 - u') + \nabla_x L(s, u, u') \cdot ((\varrho/2)w + v_0 - u)$$
$$\leq L\left(s, (\varrho/2)w + v_0, (\varrho/2)w' + v'_0\right) - L(s, u, u')$$
$$+ \left[1 + |L(s, u, u')| + \left|L\left(s, (\varrho/2)w + v_0, (\varrho/2)w' + v'_0\right)\right|\right],$$

whence

$$\left[\nabla_\xi L(s, u, u') \cdot w' + \nabla_x L(s, u, u') \cdot w\right]^+ \in L^1(s_0, s_1).$$

By the arbitariness of w, it follows that

$$\nabla_\xi L(s, u, u') \in L^1(s_0, s_1; \mathbb{R}^N), \qquad \nabla_x L(s, u, u') \in L^1(s_0, s_1; \mathbb{R}^N).$$

If, more specifically, $w \in C^1_c(]s_0, s_1[; \mathbb{R}^N)$ with $\|w\|_\infty \leq 1$, then we have

$$\int_{s_0}^{s_1} \left[\nabla_\xi L(s, u, u') \cdot ((\varrho/2)w' + v'_0 - u') + \nabla_x L(s, u, u') \cdot ((\varrho/2)w + v_0 - u)\right] ds$$
$$\geq \langle \mu, (\varrho/2)w + v_0 - u \rangle,$$

whence

$$\int_{s_0}^{s_1} \left[\nabla_\xi L(s, u, u') \cdot w' + \nabla_x L(s, u, u') \cdot w\right] ds - \langle \mu, w \rangle$$
$$\geq \frac{2}{\varrho}\left\{\int_{s_0}^{s_1} \left[\nabla_\xi L(s, u, u') \cdot (u' - v'_0) + \nabla_x L(s, u, u') \cdot (u - v_0)\right] ds \right.$$
$$\left. - \langle \mu, u - v_0 \rangle\right\}.$$

It follows that $\nabla_\xi L(s, u, u') \in BV(s_0, s_1; \mathbb{R}^N) \subseteq L^\infty(s_0, s_1; \mathbb{R}^N)$.

Let now (v_k) be a sequence in $C^1([s_0, s_1]; \mathbb{R}^N)$ converging to u in $W^{1,1}(s_0, s_1; \mathbb{R}^N)$ with $v_k(s_0) = u_0$ and $v_k(s_1) = u_1$. We have

$$\int_{s_0}^{s_1} \left[\nabla_\xi L(s, u, u') \cdot ((\varrho/2)w' + v'_k - u') + \nabla_x L(s, u, u') \cdot ((\varrho/2)w + v_k - u)\right] ds$$
$$\geq \langle \mu, (\varrho/2)w + v_k - u \rangle$$

eventually as $k \to \infty$. Going to the limit as $k \to \infty$, we get

$$\int_{s_0}^{s_1} \left[\nabla_\xi L(s, u, u') \cdot w' + \nabla_x L(s, u, u') \cdot w \right] ds \geq \langle \mu, w \rangle$$

$$\text{for all } w \in C_c^1(]s_0, s_1[; \mathbb{R}^N) \text{ with } \|w\|_\infty \leq 1 \,,$$

whence

$$\int_{s_0}^{s_1} \left[\nabla_\xi L(s, u, u') \cdot w' + \nabla_x L(s, u, u') \cdot w \right] ds = \langle \mu, w \rangle \qquad \text{for all } w \in C_c^1(]s_0, s_1[; \mathbb{R}^N)$$

and assertion (b) follows by Proposition 4.4.5. $\square$

Corollary 4.5.3 *Let $u \in W^{1,1}(s_0, s_1; \mathbb{R}^N)$ with $f(u) < +\infty$ and $0 \in \partial f(u)$. Then u is a solution of (P). If, furthermore, L is of class C^1 on $[s_0, s_1] \times \mathbb{R}^N \times \mathbb{R}^N$, then we also have*

$$\begin{cases} \left[L(s, u, u') - u' \cdot \nabla_\xi L(s, u, u') \right] \in C^1([s_0, s_1]) \,, \\ \left[L(s, u, u') - u' \cdot \nabla_\xi L(s, u, u') \right]' = D_s L(s, u, u') \qquad in [s_0, s_1] \,. \end{cases}$$

Proof It follows from assertion (b) of Theorem 4.5.2 and Proposition 4.4.5. $\square$

Corollary 4.5.4 *Let $u \in W^{1,1}(s_0, s_1; \mathbb{R}^N)$ be a local minimum of f. Then $f(u) < +\infty$ and $0 \in \partial f(u)$. In particular, the assertion of Corollary 4.5.3 holds.*

Proof Since $f(v) < +\infty$ whenever $v \in C^1([s_0, s_1]; \mathbb{R}^N)$ with $v(s_0) = u_0$ and $v(s_1) = u_1$, we have $f(u) < +\infty$. Then the assertion follows from Remark 4.3.3 and Theorem 4.3.19. $\square$

Theorem 4.5.5 *Let $(u, \lambda) \in \text{epi}(f)$ with $f(u) < \lambda$. Then $\left(\overline{\mathcal{G}_f} \right)^0 ((u, \lambda); (0, -1)) = -1$. In particular, we have $\left| d\overline{\mathcal{G}_f} \right| (u, \lambda) = \left| d\mathcal{G}_f \right| (u, \lambda) = 1$ and, for every $c \in \mathbb{R}$, the function f satisfies condition $(epi)_c$.*

Proof Since $\mathcal{G}_f$ is Lipschitz continuous of constant 1, it is easily seen that

$$\left(\overline{\mathcal{G}_f} \right)^0 ((u, \lambda); (0, -1)) \geq -1 \,.$$

To prove the opposite inequality, it is equivalent to show that

$$\left(\overline{\mathcal{G}_f} \right)^0 ((u, \lambda); (0, f(u) - \lambda)) \leq f(u) - \lambda \,, \tag{4.7}$$

as $\left(\overline{\mathcal{G}_f} \right)^0 ((u, \lambda); \cdot)$ is positively homogeneous of degree 1.

Let $R \geq 1 + \|u\|_\infty$. As before, we may assume without loss of generality that $L(s, x, \xi) \geq 0$ whenever $|x| \leq R$.

Given $\varepsilon > 0$, let $\varepsilon' > 0$ be such that

$$\varepsilon' \left(2 + \lambda + f(u)\right) \leq \frac{\varepsilon}{4}$$

and let $\delta_{R,\varepsilon'}$ be as in assumption (L_3). If $\sigma > 0$ satisfies

$$\sigma \leq 1, \quad \sigma \leq \delta_{R,\varepsilon'}, \quad \sigma \leq \frac{\varepsilon}{4},$$

we have $\|z\|_\infty \leq R$ and $\|z - u\|_\infty \leq \delta_{R,\varepsilon'}$ whenever $\|z - u\|_\infty \leq \sigma$. From assumption (L_3) it follows that

$$
\begin{aligned}
f(z + t(u - z)) &\leq f(z) + t(f(u) - f(z)) + \varepsilon' t \left(1 + f(z) + f(u)\right) \\
&\leq \eta + t(f(u) - \eta) + \varepsilon' t \left(1 + \eta + f(u)\right) \\
&\leq \eta + t\left[f(u) - \lambda + \sigma + \varepsilon' \left(1 + \lambda + \sigma + f(u)\right)\right] \\
&\leq \eta + t\left[f(u) - \lambda + \frac{\varepsilon}{2}\right],
\end{aligned}
$$

for all $(z, \eta) \in B_\sigma(u, \lambda) \cap \operatorname{epi}(f)$ and $t \in [0, \sigma]$.

In particular, if we set

$$\mathcal{V}((z, \eta), t) = \left(u - z, f(u) - \lambda + \frac{\varepsilon}{2}\right)$$

for all $(z, \eta) \in B_\sigma(u, \lambda) \cap \operatorname{epi}(f)$ and $t \in]0, \sigma]$, we have

$$\|\mathcal{V}((z, \eta), t) - (0, f(u) - \lambda)\| < \varepsilon,$$

$$(z, \eta) + t\mathcal{V}((z, \eta), t) \in \operatorname{epi}(f),$$

$$\overline{\mathcal{G}_f}((z, \eta) + t\mathcal{V}((z, \eta), t)) = \overline{\mathcal{G}_f}(z, \eta) + t\left[f(u) - \lambda + \frac{\varepsilon}{2}\right],$$

whence by Proposition 4.3.20

$$(\overline{\mathcal{G}_f})^0_\varepsilon((u, \lambda); (0, f(u) - \lambda)) \leq f(u) - \lambda + \frac{\varepsilon}{2}.$$

Going to the limit as $\varepsilon \to 0$, formula (4.7) follows, whence

$$\left(\overline{\mathcal{G}_f}\right)^0((u, \lambda); (0, -1)) = -1.$$

If $\mu \in \partial\overline{\mathcal{G}_f}(u, \lambda)$, we have

$$-\|\mu\| \leq \langle \mu, (0, -1)\rangle \leq \left(\overline{\mathcal{G}_f}\right)^0((u, \lambda); (0, -1)) = -1,$$

whence $\|\mu\| \geq 1$ for all $\mu \in \partial\overline{\mathcal{G}_f}(u, \lambda)$.

From Theorem 4.3.19 we infer that $\left|d\overline{\mathcal{G}_f}\right|(u,\lambda) = \left|d\mathcal{G}_f\right|(u,\lambda) = 1$. In particular, for every $c \in \mathbb{R}$, the function f satisfies condition $(epi)_c$.   □

4.6 A Coercive Case

Throughout this section, we assume that L satisfies assumptions (L_1) and (L_3). Moreover, instead of (L_2), we require a global coercivity, namely that:

(L_4) *there exist $C^{(0)} \geq 0$ and, for every $M > 0$, $C_M^{(1)} \geq 0$ such that*

$$L(s, x, \xi) \geq M|\xi| - C_M^{(1)} - C^{(0)}|x|$$

for all $s \in [s_0, s_1]$ and $x, \xi \in \mathbb{R}^N$.

Of course, assumption (L_4) implies (L_2).

Proposition 4.6.1 *For every $c \in \mathbb{R}$, there exists $R > 0$ such that*

$$\|u\|_\infty \leq R \qquad \text{for all } u \in W^{1,1}(s_0, s_1; \mathbb{R}^N) \text{ with } f(u) \leq c.$$

Proof Let $C^{(0)}$ be as in assumption (L_4) and let $M > 0$ be such that

$$C^{(0)} \int_{s_0}^{s_1} |u|\,ds \leq \frac{M}{2}\left(|u_0| + \int_{s_0}^{s_1} |u'|\,ds\right)$$

$$\text{for all } u \in W^{1,1}(s_0, s_1; \mathbb{R}^N) \text{ with } u(s_0) = u_0 \text{ and } u(s_1) = u_1.$$

If $C_M^{(1)}$ is given by assumption (L_4), we infer that

$$f(u) \geq M \int_{s_0}^{s_1} |u'|\,ds - C_M^{(1)} - C^{(0)} \int_{s_0}^{s_1} |u|\,ds \geq \frac{M}{2}\int_{s_0}^{s_1} |u'|\,ds - C_M^{(1)} - \frac{M}{2}|u_0|$$

$$\text{for all } u \in W^{1,1}(s_0, s_1; \mathbb{R}^N) \text{ with } u(s_0) = u_0 \text{ and } u(s_1) = u_1$$

and the assertion follows.   □

Theorem 4.6.2 *The functional f admits a global minimum $u_{gm} \in W^{1,1}(s_0, s_1; \mathbb{R}^N)$. In particular, u_{gm} is a solution of (P).*

Proof Since $f(v) < +\infty$ whenever $v \in C^1([s_0, s_1]; \mathbb{R}^N)$ with $v(s_0) = u_0$ and $v(s_1) = u_1$, there exists $c \in \mathbb{R}$ such that the set

$$\left\{u \in W^{1,1}(s_0, s_1; \mathbb{R}^N) : f(u) \leq c\right\}$$

is not empty.

From Proposition 4.6.1 and Theorem 4.4.1 we infer that f admits a global minimum $u_{gm} \in W^{1,1}(s_0, s_1; \mathbb{R}^N)$. From Corollaries 4.5.4 and 4.5.3 it follows that u_{gm} is a solution of (P). □

Theorem 4.6.3 *Let $u_{lm}, u_{gm} \in W^{1,1}(s_0, s_1; \mathbb{R}^N)$ be such that u_{lm} is a local minimum of f, u_{gm} is a global minimum of f and $u_{gm} \neq u_{lm}$.*

Then there exists $u_{mp} \in W^{1,1}(s_0, s_1; \mathbb{R}^N)$ with $f(u_{mp}) < +\infty$, $0 \in \partial f(u_{mp})$, $u_{mp} \neq u_{lm}$ and $u_{mp} \neq u_{gm}$. In particular, u_{lm}, u_{gm} and u_{mp} are three distinct solutions of (P).

Proof By Corollary 4.4.2 the functional

$$f : C([s_0, s_1]; \mathbb{R}^N) \to]-\infty, +\infty]$$

is lower semicontinuous and, by Proposition 4.6.1, Corollary 4.4.3 and Theorem 4.5.5, the functional f satisfies $(PS)_c$ and $(epi)_c$ for all $c \in \mathbb{R}$.

Taking into account Corollaries 4.5.4 and 4.5.3, we have that u_{lm}, u_{gm} are two distinct solutions of (P) and $f(u_{gm}) \leq f(u_{lm}) < +\infty$. In particular, we have $u_{lm}, u_{gm} \in C^1([s_0, s_1]; \mathbb{R}^N)$.

If we set

$$\varphi(t) = (1 - t)u_{lm} + tu_{gm},$$

we have that $\varphi \in C([0, 1]; C([s_0, s_1]; \mathbb{R}^N))$ with $\varphi(0) = u_{lm}$, $\varphi(1) = u_{gm}$ and $f \circ \varphi$ bounded.

From Theorem 4.3.15 we infer that there exists $u_{mp} \in W^{1,1}(s_0, s_1; \mathbb{R}^N)$ with $f(u_{mp}) < +\infty$, $|df|(u_{mp}) = 0$, $u_{mp} \neq u_{lm}$ and $u_{mp} \neq u_{gm}$.

By Theorem 4.3.19 we have $0 \in \partial f(u_{mp})$ and by Corollary 4.5.3 u_{mp} is a further solution of (P). □

4.7 Fermat's Principle and Numerical Evaluations for Light Rays

Consider $\mathbb{R}^3$ as an isotropic, possibly nonhomogeneous, medium whose *refractive index n* is described by a function

$$n : \mathbb{R}^3 \to [1, +\infty[$$

of class C^1.

Given $u_0, u_1 \in \mathbb{R}^3$ and $t_0 \in \mathbb{R}$, a light ray, starting from u_0 at the time t_0 and reaching u_1, is described by a map $(u, \tau) \in W^{1,1}(0, 1; \mathbb{R}^3) \times W^{1,1}(0, 1)$ which makes stationary the time functional

$$T(u, \tau) = \tau(1) - t_0$$

defined on the space

$$\Xi = \Big\{ (u, \tau) \in W^{1,1}(0, 1; \mathbb{R}^3) \times W^{1,1}(0, 1) :$$

$$u(0) = u_0, \ u(1) = u_1, \ \tau(0) = t_0, \ c\,\tau'(s) = n(u(s))|u'(s)| \ \text{ for a.a. } s \in]0, 1[\Big\},$$

where c is the speed of light in vacuum (Fermat's principle, see e.g. [11]).

Because of the expression of Ξ, it is clearly equivalent to make stationary the optical length functional

$$\mathcal{L}(u) = \int_0^1 n(u(s))|u'(s)|\, ds$$

on the space

$$\Big\{ u \in W^{1,1}(0, 1; \mathbb{R}^3) : \quad u(0) = u_0, \quad u(1) = u_1 \Big\}$$

and then recover the function τ through the relation

$$\tau(s) = t_0 + \frac{1}{c} \int_0^s n(u(\sigma))|u'(\sigma)|\, d\sigma \, .$$

So far, there is no $1 - 1$ correspondence between maps (u, τ) and physical light rays, because of the invariance of $\mathcal{L}$ by change of parametrization. Such a $1 - 1$ correspondence can be obtained, as it is well known, by adding the further condition

$$n(u(s))|u'(s)| \qquad \text{is constant for a.a. } s \in]0, 1[\, .$$

At this point it is equivalent, and more comfortable, to make stationary the functional

$$\mathcal{E}(u) = \frac{1}{2} \int_0^1 [n(u(s))]^2\, |u'(s)|^2\, ds$$

defined on the space

$$\Big\{ u \in W^{1,2}(0, 1; \mathbb{R}^3) : \quad u(0) = u_0, \quad u(1) = u_1 \Big\},$$

while the further condition

$$n(u(s))|u'(s)| \qquad \text{is constant for a.a. } s \in]0, 1[$$

is automatically satisfied, if u makes $\mathcal{E}$ stationary (see also Proposition 4.4.5). In such a case, we have $\mathcal{L}(u) = \sqrt{2\,\mathcal{E}(u)}$ and the function τ is recovered as

$$\tau(s) = t_0 + \frac{\mathcal{L}(u)}{c}\, s\,.$$

In the end, if $\mathcal{L}(u) > 0$ the light ray is also described by the space-time relation

$$x(t) = u\left(\frac{c(t - t_0)}{\mathcal{L}(u)}\right),$$

where the time t ranges over the interval $\left[t_0, t_0 + \frac{\mathcal{L}(u)}{c}\right]$.

To provide a precise mathematical setting in the line of the previous sections, let us extend $\mathcal{E}$ to $C([0, 1]; \mathbb{R}^3)$ by the value $+\infty$. Then, we get the functional

$$f : C([0, 1]; \mathbb{R}^3) \to\,]-\infty, +\infty]$$

defined by

$$f(u) = \begin{cases} \dfrac{1}{2} \displaystyle\int_0^1 [n(u(s))]^2\, |u'(s)|^2\, ds & \text{if } u \in W^{1,2}(0, 1; \mathbb{R}^3),\, u(0) = u_0 \\[2mm] & \quad \text{and } u(1) = u_1\,, \\[2mm] +\infty & \text{otherwise}\,, \end{cases}$$

which is of type we have already considered with

$$L(s, x, \xi) = \frac{1}{2} [n(x)]^2\, |\xi|^2\,, \qquad \nabla_\xi L(s, x, \xi) = [n(x)]^2\, \xi\,, \qquad \nabla_x L(s, x, \xi) = n(x)\, |\xi|^2\, \nabla n(x)\,.$$

Since n is of class C^1, the assumptions (L_1), (L_3) and (L_4) are satisfied (see also Proposition 4.5.1). A fortiori, condition (L_2) holds. In particular, the results of Sects. 4.5 and 4.6 apply.

If u is a solution of (P), we have

$$[n(u)]^2\, u' = \nabla_\xi L(s, u, u') \in C^1([0, 1]; \mathbb{R}^3)$$

whence, in this case, $u \in C^2([0, 1]; \mathbb{R}^3)$ and

$$[n(u)]^2\, u'' + 2n(u)\, (\nabla n(u) \cdot u')\, u' = \left\{[n(u)]^2\, u'\right\}' = n(u)\, |u'|^2\, \nabla n(u)\,,$$

namely

$$u''(s) = -2\left(\frac{\nabla n(u(s))}{n(u(s))} \cdot u'(s)\right) u'(s) + |u'(s)|^2\, \frac{\nabla n(u(s))}{n(u(s))}\,.$$

Moreover, since

$$L(s, u, u') - u' \cdot \nabla_\xi L(s, u, u') = -\frac{1}{2} [n(u)]^2 |u'|^2 ,$$

from Proposition 4.4.5 we infer that the function

$$\{s \mapsto n(u(s))|u'(s)|\}$$

is constant.

As an example, let us consider a two dimensional case. More precisely, let $N = 2$ and let

$$n(x) = \left(\frac{5}{2} - \frac{3}{\pi} \arctan\left(50 x^{(2)}\right)\right)^{\frac{1}{2}} , \qquad x = \left(x^{(1)}, x^{(2)}\right) .$$

This choice of n corresponds to a "regularization" of a discontinuous refractive index which takes the value 2 where $x^{(2)} < 0$ and the value 1 where $x^{(2)} \geq 0$ (Fig. 4.4). Let us refer the reader to [2, 21] for a study of the discontinuous case on Riemannian manifolds.

We aim to study the case in which $u_0 = (-d, -2)$ and $u_1 = (d, -2)$ with $d > 0$. Since $n(x)$ is almost constant when $x^{(2)}$ is close to -2, we may expect that the functional f admits a local minimum u_{lm} with

$$u_{lm}(s) \approx (1 - s) u_0 + s u_1 .$$

A more precise evaluation of the local minimum u_{lm} can be obtained by a steepest descent method (see e.g. [8, Section 3]), starting from

$$(1 - s) u_0 + s u_1 .$$

However, for certain values of d, u_{lm} cannot be a global minimum, because one obtains a lower value of the functional f by means of a trajectory which increases $x^{(2)}$ to the level where n has a lower value, moves in that area and finally comes back to the level $x^{(2)} = -2$.

This is the case, if $u_0 = (-4, -2)$, $u_1 = (4, -2)$, and again, by a steepest descent method, one can obtain a numerical evaluation of the global minimum u_{gm}. An approximate description of u_{lm} and u_{gm} is given by (Fig. 4.5)

$$u'_{lm}(0) \approx (7.999969, 0.019204) \qquad f(u_{lm}) \approx 127.6942 \qquad \mathcal{L}(u_{lm}) \approx 15.9809$$

$$u'_{gm}(0) \approx (3.910207, 6.545254) \qquad f(u_{gm}) \approx 115.9823 \qquad \mathcal{L}(u_{gm}) \approx 15.2304$$

According to Theorem 4.6.3, there is a further critical point u_{mp} of f, of mountain pass type, which represents a light ray of total internal reflection. A

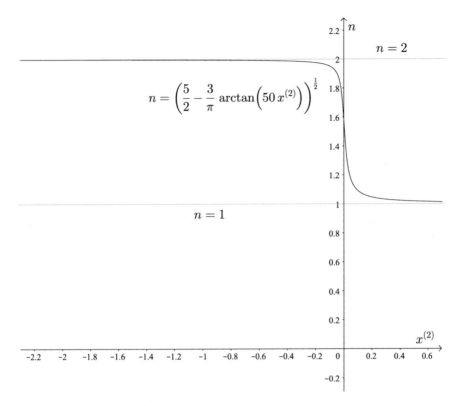

Fig. 4.4 The graph of the refractive index n

numerical evaluation of u_{mp} is more complicated, but can be performed using the technique introduced in [8] (see also [24]).

Let us collect some approximate descriptions (Fig. 4.6):

$$u_0 = (-4, -2) \qquad\qquad u_1 = (4, -2)$$

$$u'_{lm}(0) \approx (7.999969, 0.019204) \qquad f(u_{lm}) \approx 127.6942 \qquad \mathcal{L}(u_{lm}) \approx 15.9809$$

$$u'_{gm}(0) \approx (3.910207, 6.545254) \qquad f(u_{gm}) \approx 115.9823 \qquad \mathcal{L}(u_{gm}) \approx 15.2304$$

$$u'_{mp}(0) \approx (7.903412, 4.062619) \qquad f(u_{mp}) \approx 157.5607 \qquad \mathcal{L}(u_{mp}) \approx 17.7517$$

When d decreases from 4 to 3, so that u_0 and u_1 approach, for a certain value of d the local minimum close to

$$(1 - s)\, u_0 + s\, u_1$$

becomes global, so that

$$u_{gm}(s) \approx (1 - s)\, u_0 + s\, u_1\,.$$

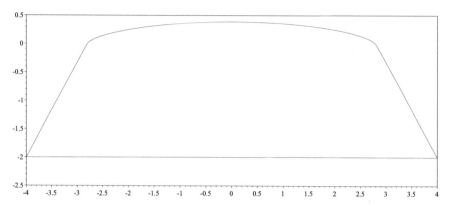

Fig. 4.5 The images of the local minimum u_{lm} (the lower curve) and of the global minimum u_{gm} (the upper curve), in the case $u_0 = (-4, -2)$ and $u_1 = (4, -2)$

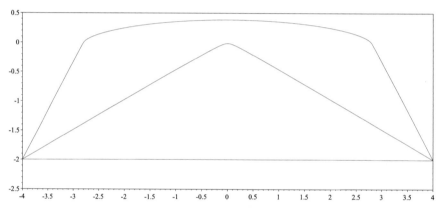

Fig. 4.6 The images of u_{lm} (the lower curve), u_{mp} (the middle curve) and u_{gm} (the upper curve), in the case $u_0 = (-4, -2)$ and $u_1 = (4, -2)$

However the old global minimum moves and becomes in fact a local minimum. Again, there is also a further critical point of mountain pass type.

This is the case if $u_0 = (-3, -2)$, $u_1 = (3, -2)$. Let us provide some approximate descriptions (Fig. 4.7):

$u_0 = (-3, -2)$ $\qquad\qquad\qquad$ $u_1 = (3, -2)$

$u'_{gm}(0) \approx (5.999987, 0.010787)$ $\qquad$ $f(u_{gm}) \approx 71.8280$ $\qquad$ $\mathcal{L}(u_{gm}) \approx 11.9857$

$u'_{lm}(0) \approx (3.410880, 5.644203)$ $\qquad$ $f(u_{lm}) \approx 86.7746$ $\qquad$ $\mathcal{L}(u_{lm}) \approx 13.1738$

$u'_{mp}(0) \approx (5.912067, 4.047692)$ $\qquad$ $f(u_{mp}) \approx 102.4277$ $\qquad$ $\mathcal{L}(u_{mp}) \approx 14.3128$

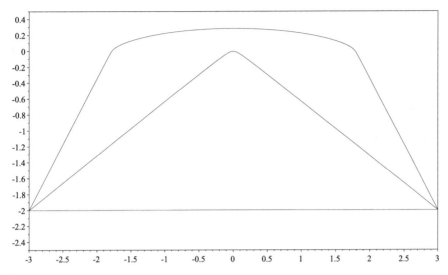

Fig. 4.7 The images of u_{gm} (the lower curve), u_{mp} (the middle curve) and u_{lm} (the upper curve), in the case $u_0 = (-3, -2)$ and $u_1 = (3, -2)$

When d is still decreasing, the local minimum and the mountain pass point approach. Let us see the situation when $u_0 = (-1.7, -2)$, $u_1 = (1.7, -2)$ and provide some approximate descriptions (Fig. 4.8):

$$u_0 = (-1.7, -2) \qquad\qquad u_1 = (1.7, -2)$$

$$u'_{gm}(0) \approx (3.399998, 0.003460) \qquad f(u_{gm}) \approx 23.0648 \qquad \mathcal{L}(u_{gm}) \approx 6.7919$$

$$u'_{lm}(0) \approx (2.868803, 4.370040) \qquad f(u_{lm}) \approx 54.5242 \qquad \mathcal{L}(u_{lm}) \approx 10.4426$$

$$u'_{mp}(0) \approx (3.239072, 4.112322) \qquad f(u_{mp}) \approx 54.6749 \qquad \mathcal{L}(u_{mp}) \approx 10.4570$$

If d is too small, hence u_0 and u_1 too close, the global minimum close to

$$(1 - s)\, u_0 + s\, u_1$$

is the unique critical point of f.

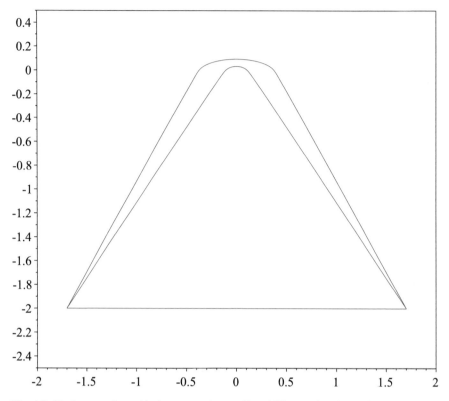

Fig. 4.8 The images of u_{gm} (the lower curve), u_{mp} (the middle curve) and u_{lm} (the upper curve), in the case $u_0 = (-1.7, -2)$ and $u_1 = (1.7, -2)$

References

1. A. Ambrosetti, P.H. Rabinowitz, Dual variational methods in critical point theory and applications. J. Funct. Anal. **14**, 349–381 (1973)
2. F. Antonacci, M. Degiovanni, On the Euler equation for minimal geodesics on Riemannian manifolds having discontinuous metrics. Discrete Contin. Dyn. Syst. **15**(3), 833–842 (2006)
3. J.M. Ball, V.J. Mizel, One-dimensional variational problems whose minimizers do not satisfy the Euler-Lagrange equation. Arch. Ration. Mech. Anal. **90**(4), 325–388 (1985)
4. L. Boccardo, T. Gallouet, Compactness of minimizing sequences. Nonlinear Anal. **137**, 213–221 (2016)
5. G. Buttazzo, M. Giaquinta, S. Hildebrandt, One-dimensional variational problems. Oxford Lecture Series in Mathematics and its Applications, vol. 15 (The Clarendon Press Oxford University Press, New York, 1998)
6. I. Campa, M. Degiovanni, Subdifferential calculus and nonsmooth critical point theory. SIAM J. Optim. **10**(4), 1020–1048 (2000)
7. G. Cerami, Un criterio di esistenza per i punti critici su varietà illimitate. Istit. Lombardo Accad. Sci. Lett. Rend. A **112**(2), 332–336 (1978)
8. Y.S. Choi, P.J. McKenna, A mountain pass method for the numerical solution of semilinear elliptic problems. Nonlinear Anal. **20**(4), 417–437 (1993)

9. F. Clarke, Optimization and nonsmooth analysis. *Canadian Mathematical Society Series of Monographs and Advanced Texts.* A Wiley-Interscience Publication (Wiley, New York, 1983)
10. F. Clarke, Functional analysis, calculus of variations and optimal control. *Graduate Texts in Mathematics*, vol. 264 (Springer, London, 2013)
11. C.J. Coleman, Point-to-point ionospheric ray tracing by a direct variational method. Radio Sci. **46**(RS5016), 1–7 (2011)
12. J.-N. Corvellec, Quantitative deformation theorems and critical point theory. Pac. J. Math. **187**(2), 263–279 (1999)
13. J.-N. Corvellec, M. Degiovanni, M. Marzocchi, Deformation properties for continuous functionals and critical point theory. Topol. Methods Nonlinear Anal. **1**(1), 151–171 (1993)
14. E. De Giorgi, A. Marino, M. Tosques, Problemi di evoluzione in spazi metrici e curve di massima pendenza. Atti Accad. Naz. Lincei Rend. Cl. Sci. Fis. Mat. Natur. (8) **68**(3), 180–187 (1980)
15. M. Degiovanni, M. Marzocchi, A critical point theory for nonsmooth functionals. Ann. Mat. Pura Appl. (4) **167**, 73–100 (1994)
16. M. Degiovanni, M. Marzocchi, On the second eigenvalue of nonlinear eigenvalue problems. Electron. J. Differ. Equ. **2018**(199), 22 pp.
17. M. Degiovanni, M. Marzocchi, Multiple critical points for symmetric functionals without upper growth condition on the principal part, in *Recent Advance in Mathematical Physics*, ed. by A. Masiello. *Symmetry* **13**(5), 898 (2021)
18. N. Dunford, J.T. Schwartz, Linear operators. Part I. *Wiley Classics Library* (Wiley, New York, 1988)
19. N. Ghoussoub, D. Preiss, A general mountain pass principle for locating and classifying critical points. Ann. Inst. H. Poincaré Anal. Non Linéaire **6**(5), 321–330 (1989)
20. A. Ioffe, E. Schwartzman, Metric critical point theory. I. Morse regularity and homotopic stability of a minimum. J. Math. Pures Appl. (9) **75**(2), 125–153 (1996)
21. R. Giambò, F. Giannoni, Minimal geodesics on manifolds with discontinuous metrics. J. Lond. Math. Soc. (2) **67**(2), 527–544 (2003)
22. G. Katriel, Mountain pass theorems and global homeomorphism theorems. Ann. Inst. H. Poincaré Anal. Non Linéaire **11**(2), 189–209 (1994)
23. J. Mawhin, M. Willem, Critical point theory and Hamiltonian systems. *Applied Mathematical Sciences*, vol. 74 (Springer, New York, 1989)
24. L. Montoro, On the numerical computation of mountain pass solutions to some perturbed semilinear elliptic problem. SEMA J. **54**, 65–90 (2011)
25. P. Pucci, J. Serrin, A mountain pass theorem. J. Differ. Equ. **60**(1), 142–149 (1985)
26. P.H. Rabinowitz, Minimax methods in critical point theory with applications to differential equations. *CBMS Regional Conference Series in Mathematics*, vol. 65, Published for the Conference Board of the Mathematical Sciences, Washington, 1986
27. M. Struwe, Variational methods. *Ergebnisse der Mathematik und ihrer Grenzgebiete*, vol. 34 (Springer, Berlin, 2008)
28. A. Visintin, Strong convergence results related to strict convexity. Commun. Partial Differ. Equ. **9**(5), 439–466 (1984)

Chapter 5
Electromagnetic Hypogene Co-seismic Sources

Giovanni Franzina

Abstract We survey some mathematical models for electro-magnetic emission due to electro-mechanically generated sources in heterogeneous materials. Because of the applications in geophysics, we focus our attention on parabolic approximations of Maxwell's equations; also, we estimate under various assumptions the discrepancy with respect to the complete set of classical electrodynamics. Then, we introduce a related inverse problem.

5.1 Introduction

The task of predicting time, location, and energy, of a seismic event in advance enough to deliver warnings is a tough task to undertake. Deterministic predictions are out-of-reach. Some estimation of earthquake probabilities is however possible. Long-term chances that an earthquake of magnitude larger than a threshold will (or will not) take place in a given area within a given time period are sometimes provided by research groups. For example, magnitude and location of the 2004 Parkfield earthquake were correctly predicted [7], without however a precise prediction of its time of occurrence; the same section of the San Andreas fault is still the object of long-term probabilistic prediction [20]. Besides the specific example, in general this kind of prediction usually regards a rather large time window, and short-term precursors are very difficult to detect.

Instead, the advance notice that a shock is going to hit an area, within several to tens of seconds, does not only involve inferential statistics, but also physics. Is called *early warning* and is sometimes made possibile by the difference in propagation velocity between primary waves and secondary waves: for instance, seismic wave

G. Franzina (✉)
Istituto Nazionale per le Applicazioni del Calcolo "M. Picone", Roma, Italy
e-mail: giovanni.franzina@cnr.it

M. Chiappini, V. Vespri (eds.), *Applied Mathematical Problems in Geophysics*,
C.I.M.E. Foundation Subseries 2308,
https://doi.org/10.1007/978-3-031-05321-4_5

propagation in elastic homogeneous isotropic media is described by a vector-valued displacement function u obeying the PDEs

$$\varrho \frac{\partial^2 u}{\partial t^2} = (\lambda + 2\mu)\nabla(\nabla \cdot u) - \mu \nabla \times \nabla \times u,$$

where λ, μ are Lamé's coefficients [23]. Recalling the vector calculus identity

$$\nabla(\nabla \cdot u) - \nabla \times \nabla \times u = \nabla^2 u,$$

where ∇^2 is the componentwise Laplace operator, we deduce that $\nabla \cdot u$ and $\nabla \times u$ solve wave equations with propagation speed respectively given by

$$\sqrt{\frac{\lambda + 2\mu}{\varrho}} \quad \text{and} \quad \sqrt{\frac{\mu}{\varrho}},$$

whose ratio is $\sqrt{2}$ at least. The earlier arrival of "P-waves", oscillating parallel to propagation, can be used to warn in advance of the imminent effect of the "S-waves", travelling slightly more slowly along the same direction but shaking in the orthogonal directions, and being therefore likely to carry a larger energy.

It makes sense to wonder if other "imminent" seismic precursors exist: In particular, if it makes sense to seek electromagnetic seismicity-related signals that might provide advance notice of an imminent earthquake: the possibility of links between subsurface electric currents and earthquake physics have been investigated for occasionally in the literature. Yet, little is known, nonetheless.

Here, we present an elementary magneto-quasistatic model aiming to study hypogene co-seismic source reconstruction starting from subsurface measurements of magnetic signals. There is no pretence of originality, nor is this paper supplemented by data or any specific material.

5.1.1 Magnetic Anomalies of Possible Coseismic Nature

Telluric electric currents, flowing throughout Earth's crust, can be measured; in particular, it is sometimes conjectured that those acting as sources for signals with frequency ranging between 10^{-3} and 10^3 Hz may admit some relation to seismology [19]. Possible meaningful causal explanations are indicated in friction and piezoelectric effects within rocks, due to the relative movement of fault blocks.

Measurements in experimental seismo-electromagnetic research were carried out to figure out about evidences of aperiodic changes in electromagnetic fields, with controversial results; in fact, claims of magnetic anomalies in the low-frequency band are sometimes asserted. The first instances in literature of papers supporting with some data these hypothesises seem to be those concerning two seismic events, one in Spitak, in Armenia, and another one in Loma Prieta, in California in 1989; an anomalous electromagnetic emission in the ULF range was measured in both cases; the instruments were believed to reveal a transient signal for hours before and after the earthquake [13]. As a matter of fact, other scientists [24] refuse to acknowledge significance to the findings about Loma Prieta earthquake, suggesting that they would be due to a sensor malfunction. Also, a recent surge in research in geophysics in this topic is related to the DEMETER mission; in this case, the magnetic measurements are taken in orbit rather than at the surface; some authors related their findings to earthquakes in Sichuan [25] and Haiti [6].

Thenceforth, a number of experts started investigating the matter, with the object of understanding if the simultaneous occurrence of seismic activity in the crust and of electromagnetic anomalies in ULF bands does take place, and if this happens by chance, with no cause-effect relation, or if instead the two phaenomena are linked by a causal relationship [9, 12, 15, 17]. For instance, in [17] the authors conjecture a source-generating mechanism based on micro-crack propagation. Their considerations are based on a simple dimensional analysis. According to findings based on geometric deep sounding, the macroscopic crustal *dielectric permittivity* ε is small relative to the average *conductivity* σ of rocks: precisely, the ratio ε/σ is estimated to range between 10^{-7} s and 10^{-5} s. Since changes in geomagnetic fields, geoelectric potentials, and electrokinetic potential on the water-solid contact are not expected to cause fast ULF variation, a possible stress-induced mechanism with this time-scale is the opening of cracks with lengthscale between 10^{-4} and 10^{-1} m at the seismic velocity of 10^{3} m/s. In this picture, the consequent EM noise would dissipate within the region interested by the phaenomenon, producing ULF emission under a cut-off at 1 Hz.

5.1.2 General Electrodynamic Models in Seismology

At the occurrence of a seismic event, and in correspondence with its preparatory phase, the scalar parameters (azimuth, dip and depth) describing the (affine) fault plane, the local crustal strain, and the width of the portion of plane interested by yield stress, cope with ground motion, friction and crack opening. The induced movement of electrically charged particles generates an electromagnetic signal. Subsurface charge motions take place along the field lines for a vector field v. By charge conservation, $J = \rho^{-1} v$ compensates the rate of change in time of the electric charge distribution in Earth's crust denoted by $\rho(\cdot, t)$ at time t, so that

$$\frac{\partial \rho}{\partial t} + \nabla \cdot J = 0. \tag{5.1}$$

By causality, $\boldsymbol{J}$ must be a non-stationary function; this yields a time-varying magnetic field. Indeed, (5.1) and Gauß's law $\nabla \cdot \boldsymbol{D} = \rho$ for the electric displacement field $\boldsymbol{D}$ imply (up to harmonic fields)

$$- \boldsymbol{D} + \nabla \times \boldsymbol{H} = \boldsymbol{J} \, . \tag{5.2a}$$

Also, in view of Faraday's induction law, a non-stationary magnetic induction field $\boldsymbol{B}$ induces a non-conservative electric field $\boldsymbol{E}$ satisfying

$$\frac{\partial \boldsymbol{B}}{\partial t} + \nabla \times \boldsymbol{E} = 0 \, . \tag{5.2b}$$

In addition to (5.2a) and (5.2b), Maxwell's equations include Gauss's law

$$\nabla \cdot \boldsymbol{D} = \rho \, , \tag{5.2c}$$

and the constraint of absence of magnetic sources:

$$\nabla \cdot \boldsymbol{B} = 0 \, . \tag{5.2d}$$

5.1.3 Constitutive Properties of the Propagation Medium

The medium filling Earth's crust is described by the constitutive relations

$$\boldsymbol{D} = \varepsilon \boldsymbol{E} \, , \quad \boldsymbol{B} = \mu \boldsymbol{H} \, , \quad \boldsymbol{J} = \boldsymbol{J}_0 + \sigma \boldsymbol{E} \, . \tag{5.3}$$

where ε is the *electric permittivity* $\varepsilon_0 = 8.8541878128(13) \times 10^{-12}$ F·m^{-1} and μ is the *magnetic permeability* $\mu_0 = 4\pi \times 10^{-7}$ H·m^{-1} in vacuum. The last equation in (5.3) includes a vector field $\boldsymbol{J}_0$, interpreted as the *source*, concentrated in region several to tens of kilometers deep and generating the EM signal, and the induced volume *eddy currents*, that depend on the crust stratification according to a linear Ohm-type law: the *electric conductivity* σ is a known piecewise scalar function [19], but anisotropic media can also be considered (in that case σ is tensor-valued). Inserting the constitutive laws (5.3) in the complete set of Maxwell's equations (5.2), we arrive at the system of equations

$$\nabla \times \boldsymbol{H} - \left(\sigma \boldsymbol{E} + \varepsilon \frac{\partial \boldsymbol{E}}{\partial t} \right) = \boldsymbol{J}_0$$
$$\nabla \times \boldsymbol{E} + \mu \frac{\partial \boldsymbol{H}}{\partial t} = 0 \, , \tag{5.4a}$$

subject to the differential constraint

$$\nabla \cdot (\mu \boldsymbol{H}) = 0 \, . \tag{5.4b}$$

Initial conditions, in this model, must be imposed both on the electric and on the magnetic field

$$E(0) = E_{0\varepsilon} \tag{5.5a}$$

$$H(0) = H_{0\varepsilon} \tag{5.5b}$$

with $E_{0\varepsilon}$ and $H_{0\varepsilon}$ being given, and $\nabla \cdot (\mu H_{0\varepsilon}) = 0$.

5.1.4 Grounds for Magneto-Quasistatic Models

In time-harmonic regime, the collection of all bulk terms between round brackets in the first equation of the system (5.4a) is given by multiplication of the electric field (in frequency domain) by the complex tensor

$$\sigma + i\omega\varepsilon .$$

Assuming smallness of the complex modulus $|\omega|$ would be consistent with findings of magnetic anomalies in ULF band. Also, the smallness of the time-scale $\frac{\varepsilon}{\sigma}$ (see Sect. 5.1.1) makes ε negligible, relative to σ. Thus, the interest in *ULF magnetic anomalies* in signals due to *hypogene* sources suggests one to consider the magneto-quasistatic model

$$\nabla \times H - \sigma E = J_0 \tag{5.6a}$$

$$\nabla \times E + \mu \frac{\partial H}{\partial t} = 0 , \tag{5.6b}$$

with the constraint

$$\nabla \cdot (\mu H) = 0 . \tag{5.6c}$$

In this case, we may provide an initial condition for the sole magnetic field, requiring that

$$H(0) = H_0 , \tag{5.7}$$

where the vector field H_0 is given and satisfies the compatibility condition $\nabla \cdot (\mu H_0) = 0$. Some comments on the singular limit as $\varepsilon \to 0^+$, in passing from (5.4) to (5.6), are made in Sect. 5.4. The loss of an initial condition may induce a boundary layer problem in time.

We point out that the limit problem (5.6) is or parabolic type. Indeed, multiplying (5.6a) by σ^{-1} and using the result to cancel the electric field from (5.6b) we arrive at the equation

$$\mu \frac{\partial H}{\partial t} + \nabla \times \left(\sigma^{-1} \nabla \times H \right) = \nabla \times \left(\sigma^{-1} J_0 \right). \tag{5.8}$$

This equation involves the differential operator $H \mapsto \nabla \times \left(\sigma^{-1} \nabla \times H \right)$, that is of elliptic type under natural assumptions on the coefficients (see Sect. 5.2.4).

5.1.5   Boundary Conditions

Let $\Omega \subset \mathbb{R}^3$ be an open region with smooth boundaries, filled with a medium having permittivity ε, magnetic permeability μ, and electric conductivity σ, with appropriate assumptions on μ and σ being in force (we postpone the details to Sect. 5.2.4). In order to match the degrees of freedom in components of J_0, E, H, both (5.4) and (5.6) list a number of equations that lack two scalar conditions. A convenient choice is to impose the boundary condition

$$H \times n = 0, \qquad \text{on } \partial\Omega \tag{5.9}$$

That, supplemented with these tangential boundary conditions, the magneto-quasistatic model (5.6) be well-posed is assured by suitable assumptions, that are discussed in Sect. 5.3.

The choice of limiting ourselves to consider homogeneous tangential boundary data is a mathematical artifice that causes no real restriction in Sects. 5.3, 5.4, and 5.5.

5.1.6   Parabolic Inverse Source Problems

Notwithstanding, it may be interesting to consider solutions attaining inhomogeneous data at the boundary. Those data may model measurements, at least in some subregion of the boundary surface; in this spirit, we point out that, under some circumstances, partial measurements are enough to recover the constitutive properties of the medium [5]. But the relevant inverse problem in this context is different, because it points to recover the source appearing in Eqs. (5.6), that we couple with (5.9), from the knowledge of suitable data θ.

In time-harmonic regime [4] there exist *non-radiating sources*, i.e., non-trivial right hand sides J_0 in (5.6) that are consistent with the homogeneous conditions (5.6). Thus, even a complete knowledge of the boundary data $H \times n = \theta$ would not be sufficient to determine J_0. Due to this ill-posedness, it is essential to subject the inverse source problem to some *a priori* assumptions on geometric and analytic structure of the source. Both in hyperbolic and in parabolic setting [1, 4], tangential boundary measurements uniquely dictate the source if the source is a priori known to be concentrated along a surface. Uniqueness holds for dipole sources, too. In Sect. 5.6, we survey this results in the time-dependent model.

Of course source reconstruction from the knowledge of tangential boundary measurements is not the only inverse source problem that can be considered, and there are a number of variants of the same idea. Another inverse problem that we introduce in Sect. 5.6 is that of determining the source in (5.6), under homogeneous boundary conditions (5.9), by assuming the complete knowledge of the magnetic field at the endpoints of a time interval [22]. Yet another example is provided by inverse source reconstruction for (5.6) with (5.9) from boundary normal measurements: the method of [21] for (5.4) can be adapted verbatim to the parabolic setting.

5.2   Mathematical Framework

Unless otherwise specified, here and henceforth the spaces of L^2 scalar-valued, vector-valued, and tensor-valued functions will be denoted by $L^2(\Omega)$, $L^2(\Omega; \mathbb{R}^3)$, and $L^2(\Omega; \mathbb{R}^{3\times 3})$, respectively. Also, we shall denote throughout the paper by $(\cdot\,,\,\cdot)_{L^2}$ and by $\|\cdot\|_{L^2}$ the scalar product and the norm in all these spaces. Occasionally, we may opt for notation $\|\cdot\|_{L^2(f)}$ when referring to the weighted L^2-norm with the function f as a density.

5.2.1   Energy Space

We recall that

$$H^1(\mathrm{curl}\,,\,\Omega) = \{\boldsymbol{\psi} \in L^2(\Omega; \mathbb{R}^3) : \nabla \times \boldsymbol{\psi} \in L^2(\Omega; \mathbb{R}^3)\}$$

is a Hilbert space with the scalar product defined for all $\boldsymbol{\varphi}, \boldsymbol{\psi}$ by $(\boldsymbol{\varphi}\,,\,\boldsymbol{\psi})_{L^2} + (\nabla \times \boldsymbol{\varphi}\,,\,\nabla \times \boldsymbol{\psi})_{L^2}$. For all smooth surfaces Σ, in particular for all smooth portions of $\partial\Omega$, we set

$$H^{-\frac{1}{2}}(\mathrm{div}_\tau;\,\Sigma) = \left\{\boldsymbol{\lambda} \in H^{-\frac{1}{2}}(\Sigma\,;\,\mathbb{R}^3) : \boldsymbol{\lambda} \cdot \boldsymbol{n} = 0,\ \mathrm{div}_\tau \boldsymbol{\lambda} = 0\right\}. \tag{5.10}$$

We recall that the Gauss Green-type formula

$$(\boldsymbol{\varphi}\,,\,\nabla \times \boldsymbol{\psi})_{L^2} - (\boldsymbol{\psi}\,,\,\nabla \times \boldsymbol{\varphi})_{L^2} = \int_{\partial\Omega} \boldsymbol{\varphi} \cdot (\boldsymbol{n} \times \boldsymbol{\psi})\,dS, \tag{5.11}$$

holds for all $\boldsymbol{\varphi}, \boldsymbol{\psi} \in C^1(\overline{\Omega}; \mathbb{R}^3)$. As a consequence, the tangential trace $\boldsymbol{\psi} \longmapsto \boldsymbol{n} \times \boldsymbol{\psi}$ from $C^1(\overline{\Omega}; \mathbb{R}^3)$ to $C(\partial\Omega; \mathbb{R}^3)$ extends to a bounded linear operator from $H^1(\mathrm{curl}, \Omega)$ to the dual space $H^{-\frac{1}{2}}(\partial\Omega; \mathbb{R}^3)$ of $H^{\frac{1}{2}}(\partial\Omega; \mathbb{R}^3)$ (see, e.g., [10]), whose kernel is denoted by $H_0^1(\mathrm{curl}, \Omega)$.

We shall occasionally abbreviate $H^1(\mathrm{curl}, \Omega)$ to $\mathscr{H}^1$, for ease of notation. We also set

$$\mathscr{H}_0^1 = \{\boldsymbol{\psi} \in H_0^1(\mathrm{curl}, \Omega) : \nabla \cdot (\mu\boldsymbol{\psi}) = 0\}, \qquad (5.12)$$

which defines a closed vector subspace of $\mathscr{H}^1$. In view of assumptions made in Sect. 5.1.5 the open set Ω supports the so-called Gaffney inequality [8] (a Friedrichs-Poincaré type functional inequality)

$$\int_\Omega |\boldsymbol{\psi}|^2 \, d\boldsymbol{x} \le C \int_\Omega |\nabla \times \boldsymbol{\psi}|^2 \, d\boldsymbol{x}, \qquad \text{for all } \boldsymbol{\psi} \in \mathscr{H}_0^1,$$

and $\mathscr{H}_0^1$ is a subspace of $H^1(\Omega; \mathbb{R}^3)$, and with the equivalent norm $\boldsymbol{\psi} \mapsto \|\nabla \times \boldsymbol{\psi}\|_{L^2}$, and $\mathscr{H}_0^1$ is contained in $L^2(\Omega; \mathbb{R}^3)$ with a compact embedding (see, e.g., [14, §2]). By induction, we also define

$$\mathscr{H}_0^n = \{\boldsymbol{\psi} \in \mathscr{H}_0^{n-1} : \nabla \times \boldsymbol{\psi} \in \mathscr{H}_0^{n-1}, \nabla \cdot \boldsymbol{\psi} = 0\}$$

for $n \in \{1, 2, 3\}$.

5.2.2 Time-Dependent Spaces

We recall that, given $p \ge 1$ and a Hilbert space $\mathcal{Z}$, a function ϕ belongs to $L^p(0, T; \mathcal{Z})$ if we have

$$\int_0^T \|\phi(t)\|_{\mathcal{Z}}^p \, dt < +\infty$$

and, in that case, the p-th root of left hand side is denoted by $\|\phi\|_{L^p(0,T;\mathcal{Z})}$. We also recall that this defines a complete norm on $L^p(0, T; \mathcal{Z})$. The same conclusion holds in the borderline case $p = \infty$, provided that the p-th root of the integral is replaced by the essential sup norm.

If $\partial_t\phi$ belongs to $L^p(0, T; \mathcal{Z})$ then so does ϕ, and in that case we write $\phi \in W^{1,p}(0, T; \mathcal{Z})$. Assume that $\mathcal{Z} \subset L^2(\Omega; \mathbb{R}^3) \subset \mathcal{Z}^*$, where $\mathcal{Z}^*$ is the dual of $\mathcal{Z}$, with continuous inclusions having dense images. Then, we recall Lions-Magenes Lemma (see [11, §5]): if $\phi \in L^2(0, T; \mathcal{Z})$ and $\partial_t\phi$ belongs to the dual space $L^2(0, T; \mathcal{Z}^*)$, then $\phi \in C([0, T]; L^2(\Omega; \mathbb{R}^3))$ and the function $t \mapsto \|\phi(t)\|_{L^2}^2$ is absolutely continuous, with $\frac{1}{2}\frac{d}{dt}\|\phi(t)\|_{L^2}^2 = \langle \phi(t), \partial_t\phi(t) \rangle$ for a.e. $0 \le t \le T$, where $\langle \cdot , \cdot \rangle$ denotes the duality pairing. The same conclusions hold also if $\phi \in W^{1,2}(0, T; L^2(\Omega; \mathbb{R}^3))$, because of one-dimensional Sobolev embedding.

5.2.3   Regularity of the Boundaries

For all points $x \in \mathbb{R}^3$ and for all $r > 0$, we denote by $B_r(x)$ the ball of radius r centred at x. An open set Ω in $\mathbb{R}^3$ is said to satisfy a *uniform two-sided ball condition* if there exists a positive $r > 0$ with the property that, for every boundary point $\xi \in \partial\Omega$, there exist a ball $B_r(x)$ contained in Ω and a ball $B_r(y)$ contained in its complement, such that ξ belongs both to the closure of $B_r(x)$ and to that of $B_r(y)$. Throughout this paper, we shall make the following assumptions:

Ω is a bounded open set in $\mathbb{R}^3$. (5.13a)

Ω is either convex or it satisfies a uniform two-sided ball condition. (5.13b)

The closure $\overline{\Omega}$ of Ω has the same boundary as Ω. (5.13c)

Incidentally, we point out that, under the assumption (5.13c), condition (5.13b) is equivalent to a uniform bound on the $C^{1,1}$ constants of the functions describing locally Ω as a subgraph.

5.2.4   Assumptions on the Coefficients

We assume that σ be a bounded measurable function with values in the set of (3×3)-symmetric matrices with real coefficients such that

$$\sigma_0 |\xi|^2 \le \sigma(x)\xi \cdot \xi \le \sigma_0^{-1}|\xi|^2, \qquad \text{for a.e. } x \in \Omega, \text{ and for all } \xi \in \mathbb{R}^3, \quad (5.14a)$$

for an appropriate constant $\sigma_0 > 1$. Here, for all $\xi, \eta \in \mathbb{R}^3$ we are denoting by $\xi \cdot \eta$ the standard scalar product in $\mathbb{R}^3$. By μ we shall denote a fixed positive smooth scalar function, satisfying

$$\sigma_0 \le \inf\{\mu(x), |\nabla\mu(x)\} \le \sup\{\mu(x), |\nabla\mu(x)\} \le \sigma_0^{-1}, \quad (5.14b)$$

for all $x \in \Omega$.

5.2.5   Total Basis and Magnetic Eigenvalues

Assuming (5.13) and (5.14b), the eigenvalue-type boundary value problem

$$
\begin{cases}
\nabla \times \nabla \times \boldsymbol{\psi} = \lambda \boldsymbol{\psi}, & \text{in } \Omega, \\
\nabla \cdot (\mu \boldsymbol{\psi}) = 0, & \text{in } \Omega, \\
\boldsymbol{n} \times \boldsymbol{\psi} = 0, & \text{on } \partial\Omega
\end{cases}
\tag{5.15}
$$

admits non-trivial (weak) solutions for a discrete set of real numbers λ, called *eigenvalues*. If $\lambda \geq 0$ is an eigenvalue, any non-trivial (weak) solution of (5.15) is called an *eigenfield*. The eigenvalues form an unbounded non-decreasing sequence, that is completely described by the variational principle

$$
\lambda_m = \min_{\boldsymbol{\varphi}_1,\dots,\boldsymbol{\varphi}_m \in \mathscr{H}_0^1} \max \left\{ \frac{\displaystyle\int_\Omega |\nabla \times \boldsymbol{\psi}(x)|^2 \, dx}{\displaystyle\int_\Omega |\boldsymbol{\psi}(x)|^2 \, dx} : \boldsymbol{\psi} \in \mathrm{Span}\{\boldsymbol{\varphi}_1,\dots,\boldsymbol{\varphi}_m\} \setminus \{0\} \right\}.
\tag{5.16}
$$

If Ω is simply connected with a connected boundary $\partial\Omega$ then $\lambda_1 > 0$. In general, $\lambda_j = 0$ for all positive integers j smaller than the number of degrees of freedom behind conditions

$$
\nabla \times \boldsymbol{\psi} = 0, \ \nabla \cdot (\mu \boldsymbol{\psi}) = 0, \ \boldsymbol{n} \times \boldsymbol{\psi}\big|_{\partial\Omega} = 0,
$$

which is however always finite (it is the second Betti number of Ω as a Euclidean manifold).

5.3   Well-Posedness for the Forward Problem

5.3.1   Parabolic Estimates

In this section we provide solutions of the quasi-static Maxwell equations, understood in the following sense.

Definition 5.3.1 Given $\boldsymbol{J}_0 \in L^2(0, T; L^2(\Omega; \mathbb{R}^3))$ and $\boldsymbol{H}_0 \in L^2(\Omega; \mathbb{R}^3)$, with $\nabla \cdot (\mu \boldsymbol{H}_0) = 0$ in Ω, we say that $\boldsymbol{H} \in L^2(0, T; \mathscr{H}_0^1)$, with $\partial_t \boldsymbol{H} \in L^2(0, T; (\mathscr{H}_0^1)^*)$ and $\boldsymbol{H}(0) = \boldsymbol{H}_0$, is a weak solution of (5.6) with the boundary conditions (5.9) if we have

$$
-\int_{T_0}^{T_1} \left(\mu \boldsymbol{H}, \frac{\partial \boldsymbol{\phi}}{\partial t} \right)_{L^2} d\tau + \int_{T_0}^{T_1} \left(\sigma^{-1} \nabla \times \boldsymbol{H}, \nabla \times \boldsymbol{\phi} \right)_{L^2} d\tau = \int_{T_0}^{T_1} \left(\sigma^{-1} \boldsymbol{J}_0, \nabla \times \boldsymbol{\phi} \right)_{L^2} d\tau,
$$

for all $\boldsymbol{\phi} \in C^\infty(\Omega \times [0, T])$ with support contained in $\Omega \times [0, T]$, for all $0 < T_0 < T_1 < T$.

Note that, under the assumptions made in Definition 5.3.1, the initial condition on weak solutions makes sense because of Lions- Magenes Lemma. If certain better regularity criteria are met, the weak solutions are solutions in the following stronger sense.

Definition 5.3.2 Given $\boldsymbol{J}_0 \in L^2(0, T; L^2(\Omega; \mathbb{R}^3))$ and $\boldsymbol{H}_0 \in \mathcal{H}_0^1$, with $\nabla \cdot (\mu \boldsymbol{H}_0) = 0$ in Ω, we say that $\boldsymbol{H} \in L^2(0, T; \mathcal{H}_0^1)$, with $\partial_t \boldsymbol{H} \in L^2(0, T; L^2(\Omega; \mathbb{R}^3))$, is a strong solution of (5.6), with boundary conditions (5.9), if

$$\left(\mu \frac{\partial \boldsymbol{H}}{\partial t}(\tau), \boldsymbol{\psi}\right)_{L^2} + \left(\sigma^{-1}\nabla \times \boldsymbol{H}(\tau), \nabla \times \boldsymbol{\psi}\right)_{L^2} = \left(\sigma^{-1}\boldsymbol{J}_0(\tau), \nabla \times \boldsymbol{\psi}\right)_{L^2},$$

$$(5.17)$$

for all $\boldsymbol{\psi} \in \mathcal{H}_0^1$ and for a.e. $0 \le \tau \le T$.

To construct solutions, we follow a specific Galerkin-type method. To do so, for all $m \in \mathbb{N}$, we denote by $\boldsymbol{\pi}_m$ the projection from $\mathcal{H}_0^1$ onto the vector space $\mathcal{H}_{0m}^1$ generated by the eigenfields associated with the eigenvalues $\lambda_1, \ldots, \lambda_m$ introduced in (5.16). Given $\boldsymbol{H}_0 \in \mathcal{H}_0^1$, the standard results for linear systems of ordinary differential equations imply the existence of a (unique) $\boldsymbol{H}_m \in C^1([0, T]; \mathcal{H}_{0m}^1)$ for which

$$\begin{cases} \nabla \times \boldsymbol{H}_m - \sigma \boldsymbol{E}_m = \boldsymbol{\pi}_m \boldsymbol{J}_0 \\ \nabla \times \boldsymbol{E}_m + \mu \dfrac{\partial \boldsymbol{H}_m}{\partial t} = 0 \end{cases} \qquad (5.18)$$

for an appropriate $\boldsymbol{E}_m \in C^1([0, T]; \mathcal{H}_{0m}^1)$, under initial conditions $\boldsymbol{H}_m(0) = \boldsymbol{\pi}_m \boldsymbol{H}_0$. Multiplying the first equation in (5.18) by $\boldsymbol{E}_m$ and the second one by $\boldsymbol{H}_m$, we arrive at the identity

$$\frac{1}{2}\frac{d}{dt}(\mu \boldsymbol{H}_m, \boldsymbol{H}_m)_{L^2} + (\sigma \boldsymbol{E}_m, \boldsymbol{E}_m)_{L^2} = -(\boldsymbol{\pi}_m \boldsymbol{J}_0, \boldsymbol{E}_m)_{L^2}. \qquad (5.19)$$

Using Cauchy Schwartz inequality and the first equation in (5.18) again, from (5.19) we deduce

$$\frac{d}{dt}(\mu \boldsymbol{H}_m, \boldsymbol{H}_m)_{L^2} + \left(\sigma^{-1}\nabla \times \boldsymbol{H}_m, \nabla \times \boldsymbol{H}_m\right)_{L^2} \le \left(\sigma^{-1}\boldsymbol{\pi}_m \boldsymbol{J}_0, \boldsymbol{\pi}_m \boldsymbol{J}_0\right)_{L^2}.$$

Integrating in time, we get

$$\sup_{\tau \in [0,T]} \| \sqrt{\mu} H_m \|_{L^2}^2 + \int_0^T \| \nabla \times H_m \|_{L^2}^2 \, d\tau \le C_1(\Lambda) \left(\| H_0 \|_{L^2}^2 + \int_0^T \| J_0 \|_{L^2}^2 \, d\tau \right).$$
(5.20)

Also, recalling that (5.15) with $\lambda = \lambda_i$ holds for an appropriate $\psi_i \in \mathcal{H}_{0m}^1$, we have

$$\| \nabla \times H_m(0) \|_{L^2}^2 = \sum_{i,j=1}^m (H_m(0), \psi_i)_{L^2} (H_m(0), \psi_j)_{L^2} (\nabla \times \psi_i, \nabla \times \psi_j)_{L^2}$$

$$= \sum_{i=1}^m \lambda_i | (H_0, \psi_i)_{L^2} |^2 = \sum_{\substack{i \le m \\ \lambda_i > 0}} \left| \left(\nabla \times H_0, \frac{\nabla \times \psi_i}{\sqrt{\lambda_i}} \right)_{L^2} \right| \le \| \nabla \times H_0 \|_{L^2}^2$$
(5.21)

where in the last passage we also used Bessel's inequality. We differentiate the first equation in (5.18) and we multiply the result by E_m, then we multiply the second equation in (5.18) by $\partial_t H_m$. Doing so, we arrive at

$$\left(\mu \frac{\partial H_m}{\partial t}, \frac{\partial H_m}{\partial t} \right)_{L^2} + \frac{1}{2} \frac{d}{dt} (\sigma E_m, E_m)_{L^2} \le (\sigma E_m, E_m)_{L^2}^{\frac{1}{2}} \left(\sigma^{-1} \pi_m \frac{\partial J_0}{\partial t}, \pi_m \frac{\partial J_0}{\partial t} \right)_{L^2}^{\frac{1}{2}}$$

where we also used Cauchy-Schwartz inequality. By a Grönwall-type argument, we infer that

$$(\sigma E_m, E_m)_{L^2} \le 2 (\sigma E_m(0), E_m(0))_{L^2} + T \int_0^T \left(\sigma^{-1} \pi_m \frac{\partial J_0}{\partial t}, \pi_m \frac{\partial J_0}{\partial t} \right)_{L^2} d\tau.$$

Using the first equation in (5.18) with $t = 0$ and recalling (5.21), after an integation in time from the last two inequalities we deduce

$$\int_0^T \left\| \sqrt{\mu} \frac{\partial H_m}{\partial t} \right\|_{L^2}^2 d\tau + \sup_{\tau \in [0,T]} \| \nabla \times H_m \|_{L^2}^2 \le C_2(\Lambda) \left(\| \nabla \times H_0 \|_{L^2}^2 + \int_0^T \left\| \frac{\partial J_0}{\partial t} \right\|_{L^2}^2 d\tau \right).$$
(5.22)

If $\nabla \times (\sigma^{-1} \nabla \times H_0) \in L^2(\Omega)$, the system (5.18) at initial time gives

$$\mu \partial_t H_m(0) = -\nabla \times (\sigma^{-1} \nabla \times H_0) + \nabla \times (\sigma^{-1} J_0(0)).$$

By (5.18) we also have

$$\left(\mu \frac{\partial^2}{\partial t^2} H_m, \psi \right)_{L^2} + \left(\sigma^{-1} \nabla \times \frac{\partial H_m}{\partial t}, \nabla \times \psi \right)_{L^2} = \left(\sigma^{-1} \pi_m \frac{\partial J_0}{\partial t}, \nabla \times \psi \right)_{L^2}, \qquad \text{for all } \psi \in \mathcal{H}_0^1.$$

Then, for $\nabla \times (\sigma^{-1} \nabla \times H_0) \in L^2(\Omega)$, choosing $\psi = \dfrac{\partial H_m}{\partial t}$ and integrating in time we arrive at

$$\sup_{\tau \in [0,T]} \left\| \frac{\partial H_m}{\partial t}(\tau) \right\|_{L^2}^2 + \int_0^T \left\| \frac{\partial (\nabla \times H_m)}{\partial t} \right\|_{L^2}^2 d\tau$$

$$\leq C_3(\Lambda) \left(\|\sigma^{-1} \nabla \times H_0\|_{\mathscr{H}_0^1} + \int_0^T \left\| \frac{\partial J_0}{\partial t} \right\|_{L^2}^2 d\tau \right).$$

$$(5.23)$$

The energy estimates obtained during the procedure imply the following fact.

Theorem 5.3.3 *Let $\Omega \subset \mathbb{R}^3$ satisfy (5.13), let $\sigma \in L^\infty(\Omega; \mathbb{R}^{3 \times 3})$ and let $\mu \in C^1(\mathbb{R}^3)$ be such that (5.14) holds. Given $J_0 \in L^2(0, T; L^2(\Omega; \mathbb{R}^3))$ and $H_0 \in L^2(\Omega)$, with $\nabla \cdot (\mu H_0) = 0$ in Ω, there exists a unique weak solution $H \in L^2(0, T; \mathscr{H}_0^1)$, with $\partial_t H \in L^2(0, T; (\mathscr{H}_0^1)^*)$, and we have*

$$\|H\|_{L^\infty(0,T;L^2(\Omega))} + \|H\|_{L^2(0,T;\mathscr{H}_0^1)} \leq c_1 \left(\|H_0\|_{L^2} + \|J_0\|_{L^2(0,T;L^2(\Omega))} \right).$$

$$(5.24)$$

If also $H_0 \in \mathscr{H}_0^1$ and $J_0 \in W^{1,2}(0, T; L^2(\Omega; \mathbb{R}^3))$, then H is a strong solution and

$$\|H\|_{W^{1,2}(0,T;L^2(\Omega))} + \|H\|_{L^\infty(0,T;\mathscr{H}_0^1)} \leq c_2 \left(\|H\|_{\mathscr{H}_0^1} + \left\| \frac{\partial J_0}{\partial t} \right\|_{L^2(0,T;L^2(\Omega))} \right).$$

$$(5.25)$$

Eventually, if in addition $\sigma^{-1} \nabla \times H_0 \in \mathscr{H}_0^1$ then

$$\|H\|_{W^{1,\infty}(0,T;L^2(\Omega))} + \|H\|_{W^{1,2}(0,T;\mathscr{H}_0^1)} \leq c_3 \left(\|\sigma^{-1} \nabla \times H_0\|_{\mathscr{H}_0^1} + \left\| \frac{\partial J_0}{\partial t} \right\|_{L^2(0,T;L^2(\Omega))} \right).$$

$$(5.26)$$

In the estimates, the constants c_1, c_2, c_3 depend on σ_0, T, and Ω, only.

Proof Uniqueness follows at once by the estimates. The existence of a field solving the weak equation, with the first estimate, is a consequence of (5.20) and of a routine compactness argument. In case $H_0 \in \mathscr{H}_0^1$, we can use (5.22), too: as a consequence, we obtain the second estimate; also, choosing a test function of the form $\phi(x, t) = \psi(x)h(t)$, with $h \in C_0^1(0, T)$, in the weaker equation and integrating by parts, we deduce the stronger equation for almost all times $t \in (0, T)$ because of the arbitrariness of h. Eventually, under the additional assumption that $\nabla \times (\sigma^{-1} \nabla \times H_0) \in \mathscr{H}_0^1$ we can use also (5.23) and that implies the last statement. $\square$

Remark 5.3.4 If the initial data satisfy the assumption $\nabla \times (\sigma^{-1}\nabla \times H_0) \in \mathcal{H}_0^1$, then by Aubin-Lions Lemma the convergence of the Galerkin method holds, at least, is in the following sense:

$$\lim_{m \to \infty} \int_0^T \|H_m(\cdot, t) - H(\cdot, t)\|_{L^2}^2 \, dt = 0$$

The case of heterogeneous media with conductivities that include discontinuities is interesting in applications. The following regularity-related result is proved in [14, Theorem 4.1]. The proof presented there is based on the method introduced in [2], which combines elliptic Campanato-type estimates and the classical De Giorgi-Nash regularity with the relevant Helmholtz decompositions.

Theorem 5.3.5 (Regularity) *Let $\Omega \subset \mathbb{R}^3$ satisfy (5.13), let $\sigma \in L^\infty(\Omega; \mathbb{R}^{3\times3})$ and let $\mu \in C^1(\mathbb{R}^3)$ be such that (5.14) holds. Then, there exists $\alpha_0 \in (0, \frac{1}{2}]$, only depending on σ_0, such that for every $\alpha \in (0, \alpha_0]$ the following holds: for every $H_0 \in C^{0,\alpha}(\overline{\Omega}; \mathbb{R}^3)$ and for every $J_0 \in L^2(0, T; C^{0,\alpha}(\overline{\Omega}; \mathbb{R}^3))$, if (E, H) is a weak solution of (5.6), then $H \in L^2(0, T; C^{0,\alpha}(\overline{\Omega}; \mathbb{R}^3))$, and we have*

$$\|H\|_{L^2(0,T;C^{0,\alpha}(\overline{\Omega};\mathbb{R}^3))} \leq C\left[\|H_0\|_{C^{0,\alpha}(\overline{\Omega};\mathbb{R}^3)} + \|H\|_{W^{1,2}(0,T;\mathcal{H}_0^1)} + \|J_0\|_{L^2(0,T;C^{0,\alpha}(\overline{\Omega};\mathbb{R}^3))}\right],$$
(5.27)

where the constant C depends on σ_0, T, and on Ω, only.

Remark 5.3.6 If the initial data satisfy also the assumption that $\sigma^{-1}\nabla \times H_0 \in \mathcal{H}_0^1$ then

$$\|H\|_{L^\infty(0,T;C^{0,\alpha}(\overline{\Omega};\mathbb{R}^3))} \leq C\left[\|H_0\|_{C^{0,\alpha}(\overline{\Omega};\mathbb{R}^3)} + \|H\|_{W^{1,\infty}(0,T;\mathcal{H}_0^1)} + \|J_0\|_{L^\infty(0,T;C^{0,\alpha}(\overline{\Omega};\mathbb{R}^3))}\right].$$

The appropriate assumptions on the conductivity coefficients σ do include the possibility of discontinuities, as said. Nonetheless, it can however be legit to consider simplified situations in which the singularities are concentrated along smooth geometric objects.

For example, a relevant situation is that of an heterogeneous isotropic medium described by a piecewise constant conductivity, jumping across a plane. In this simpler case, the magneto-quasistatic field solves a system of three scalar heat equations mildly coupled by the presence in the right hand side of the components of a given forcing term: in each level set of σ, we have

$$\mu\sigma\frac{\partial H}{\partial t} - \nabla^2 H = \nabla \times J_0.$$
(5.28)

In (5.28), ∇^2 denotes the vector-valued Laplace operator defined componentwise by

$$\nabla^2 \boldsymbol{\psi} = \begin{pmatrix} \partial_{xx}^2 f + \partial_{yy}^2 f + \partial_{zz}^2 f \\ \partial_{xx}^2 g + \partial_{yy}^2 g + \partial_{zz}^2 g \\ \partial_{xx}^2 h + \partial_{yy}^2 h + \partial_{zz}^2 h \end{pmatrix}, \qquad \text{for all } \boldsymbol{\psi} = \begin{pmatrix} f \\ g \\ h \end{pmatrix}.$$

Theorem 5.3.7 (Transmission Conditions) *Let $r > 0$, let $R > 0$, and let $e \in \mathbb{R}^3$ and let*

$$\Omega = \left\{ x \in \mathbb{R}^3 : |x \cdot e| \leq R, \ |x - (x \cdot e)e| \leq r \right\},$$

$$\Omega_+ = \left\{ x \in \Omega : x \cdot e > 0 \right\}, \qquad and \qquad \Omega_- = \left\{ x \in \Omega : x \cdot e < 0 \right\}.$$

Let $\sigma_+, \sigma_- > 0$, let $\sigma = \sigma_+ \cdot 1_{\Omega_+} + \sigma_- \cdot 1_{\Omega_-}$, and let $[\![\cdot]\!]$ denote the jump across $\Sigma = \{x \in \mathbb{R}^3 : x \cdot e = 0\}$ in the sense of traces. Then, for every $H_0 \in \mathscr{H}_0^1$ and for every $J_0 \in L^2(0, T; L^2(\Omega; \mathbb{R}^3))$, a vector field H is a weak solution of (5.6) if, and only if, it solves (5.28) in $\Omega_+ \cup \Omega_-$ and the transmission conditions

$$[\![H]\!] = 0 \tag{5.29a}$$

$$[\![e \times (\sigma^{-1} \nabla \times H)]\!] = 0 \tag{5.29b}$$

hold along Σ for a.e. $0 \leq t \leq T$.

Proof Because $H_0 \in \mathscr{H}_0^1$, a vector field H is a weak solution if and only if it is a strong one, i.e., (5.17) holds for all $\boldsymbol{\psi} \in \mathscr{H}_0^1$. In that case, choosing first test fields $\boldsymbol{\psi}$ with support either contained in Ω_+ or in Ω_- we see that equations (5.6a) and (5.6b) hold in $\Omega_+ \cup \Omega_- = \Omega \setminus \Sigma$. Combining those equations with (5.17) when the support $\boldsymbol{\psi}$ has non-empty intersection with Σ, after an integration by parts we arrive at

$$\int_\Sigma \boldsymbol{\psi} \cdot e \times \left(\mu \frac{\partial H}{\partial t} \right) dS = 0, \qquad \int_\Sigma \boldsymbol{\psi} \cdot e \times (\sigma^{-1} \nabla \times H) \, dS = 0.$$

As $\boldsymbol{\psi}$ can be any element of $\mathscr{H}_0^1$, we infer that

$$\left[\!\!\left[e \times \frac{\partial H}{\partial t} \right]\!\!\right] = 0, \qquad [\![e \times (\sigma^{-1} \nabla \times H)]\!] = 0,$$

and (5.29b) is proved. Because, by assumption, H_0 must not jump across Σ, the first identity in the latter implies $[\![e \times H]\!] = 0$. That $[\![e \cdot H]\!] = 0$ too, follows from (5.6c) by divergence theorem. This proves also (5.29a) and concludes the proof. $\quad\square$

Remark 5.3.8 When in force for solutions, condition (5.29a) is valid not just in the sense of traces, because H is continuous across Σ in view of Theorem 5.3.5. On the other hand, condition (5.29b) limits the regularity of $\boldsymbol{H}$ in the scale of Hölder spaces, and indicates that it must not be continuously differentiable: any jump in the coefficient σ must be compensated by the curl of $\boldsymbol{H}$.

5.4   Hyperbolic Estimates and Their Singular Limit

We devote this section to some comments on the asymptotic behaviour of electromagnetic fields in media with vanishing dielectricity. This issue, in the case of homogeneous media, has been already considered in literature, even for quasilinear models: we refer the interested reader to [16]. The problem in heterogeneous media does not give rise to particular additional issues, at least if the conducivity is described by a smooth function; for sake of simplicity, in the present section we will limit our attention to this simpler case.

Let $\boldsymbol{E}_\varepsilon$, $\boldsymbol{H}_\varepsilon$ solve the complete set of Maxwell equations (5.4) under the boundary conditions (5.9). From (5.4a), we arrive at the estimate

$$\int_0^T \|\boldsymbol{E}_\varepsilon\|_{L^2(\sigma)}^2 \, dt + \max_{0 \le t \le T}\left[\|\boldsymbol{E}_\varepsilon\|_{L^2(\varepsilon)}^2 + \|\boldsymbol{H}_\varepsilon\|_{L^2(\mu)}^2\right] \le C \int_0^T \|\boldsymbol{J}_0\|_{L^2}^2 \, dt + 2L_{0\varepsilon}^2 \tag{5.30}$$

where the constant C is independent of ε, and $L_{0\varepsilon} = \|\boldsymbol{E}_{0\varepsilon}\|_{L^2} + \|\boldsymbol{H}_{0\varepsilon}\|_{L^2}$ where $\boldsymbol{E}_{0\varepsilon}$ and $\boldsymbol{H}_{0\varepsilon}$ are the data involved in the initial conditions (5.5).

Differentiating in time equations (5.4a), multiplying the first one by $\partial_t \boldsymbol{E}_\varepsilon$ and the second one by $\partial_t \boldsymbol{H}_\varepsilon$, integrating by parts, and using (5.9), by a similar argument we may also get

$$\int_0^T \left\|\frac{\partial \boldsymbol{E}_\varepsilon}{\partial t}\right\|_{L^2(\sigma)}^2 \, dt + \max_{0 \le t \le T}\left[\left\|\frac{\partial \boldsymbol{E}_\varepsilon}{\partial t}\right\|_{L^2(\varepsilon)}^2 + \left\|\frac{\partial \boldsymbol{H}_\varepsilon}{\partial t}\right\|_{L^2(\mu)}^2\right] \le C \int_0^T \left\|\frac{\partial \boldsymbol{J}_0}{\partial t}\right\|_{L^2}^2 \, dt + 2M_{0\varepsilon}^2 , \tag{5.31}$$

with a constant C that is also independent of ε. In this case, the right hand side involves the quantity $M_{0\varepsilon} = \|\partial_t \boldsymbol{E}_\varepsilon(0)\|_{L^2} + \|\partial_t \boldsymbol{H}_\varepsilon(0)\|_{L^2}$, which does not depend on the data directly, but rather through the solutions.

Note that both $L_{0\varepsilon}$ and $M_{0\varepsilon}$ depend on ε. When considering the limit as $\varepsilon \to 0^+$ it is therefore natural to dictate some additional requirement on the initial data so as to provide uniform bounds for these quantities. In particular, a uniform bound for $L_{0\varepsilon}$ in (5.30) is a minimal requirement. This would be ensured, for example, by conditions

$$\boldsymbol{H}_{0\varepsilon} \in \mathscr{H}_0^n , \quad \boldsymbol{E}_{0\varepsilon} \in \mathscr{H}_0^n \tag{5.32}$$

and

$$A_0 := \limsup_{\varepsilon \to 0^+} \left(\|\boldsymbol{H}_{0\varepsilon}\|_{\mathcal{H}_0^n} + \|\boldsymbol{E}_{0\varepsilon}\|_{\mathcal{H}_0^n} \right) < +\infty. \tag{5.33}$$

In fact, if (5.32) and (5.33) hold then we further have that

$$\liminf_{\varepsilon \to 0^+} \|\boldsymbol{H}_{0\varepsilon} - \boldsymbol{H}_0\|_{\mathcal{H}_0^{n-1}} = 0, \tag{5.34}$$

for a suitable $\boldsymbol{H}_0 \in \mathcal{H}_0^n$.

Here and henceforth, we make use of a fixed $\rho \in C_0^\infty(\mathbb{R}^3)$, with $0 \le \rho \le 1$ and $\int \rho(\boldsymbol{x}) \, d\boldsymbol{x} = 1$, assumed to be even and have compact support in the unit ball, and for every $\delta \in (0, 1)$ we set $\rho_\delta(\boldsymbol{x}) = \delta^{-3} \rho(\boldsymbol{x}/\delta)$, for all $\boldsymbol{x} \in \mathbb{R}^3$. For every single $\delta > 0$ and for all (scalar, vector-, or tensor-valued) functions (or distributions) $\boldsymbol{u}$ we shall denote by

$$\boldsymbol{u} \star \rho_\delta(\boldsymbol{x}) = \int \rho_\delta(\boldsymbol{x} - \boldsymbol{y})\boldsymbol{u}(\boldsymbol{y}) \, d\boldsymbol{y}, \qquad \boldsymbol{x} \in \mathbb{R}^3,$$

the (componentwise) convolution product of $\boldsymbol{u}$ by the mollifier ρ_δ. We recall that $\boldsymbol{u} * \rho_\delta \in C^\infty(\mathbb{R}^3)$, and that $\boldsymbol{u} \in \mathcal{H}_0^n$ implies that $\boldsymbol{u} \star \rho_\delta \to \boldsymbol{u}$, as $\delta \to 0^+$, strongly in $\mathcal{H}_0^n$.

Theorem 5.4.1 (Weak Convergence) *Let $\Omega \subset \mathbb{R}^3$ satisfy (5.13), let σ be a smooth function, satisfying (5.14a), with $\|\sigma\|_{C^3(\overline{\Omega}; \mathbb{R}^{3 \times 3})} \le \sigma_0^{-1}$, and let μ be a positive constant, with $\mu > \sigma_0$. Let $n \in \{1, 2, 3\}$, let $\boldsymbol{J}_0 \in W^{1,2}(0, T; \mathcal{H}_0^n)$, and let $\boldsymbol{H}_0 \in \mathcal{H}_0^n$. For all $\varepsilon > 0$, let $(\boldsymbol{E}_{0\varepsilon}, \boldsymbol{H}_{0\varepsilon})$ satisfy (5.32) and (5.33), and let $(\boldsymbol{E}_\varepsilon, \boldsymbol{H}_\varepsilon)$ be a solution of Maxwell's equations (5.4) with boundary conditions (5.9) and initial conditions (5.5). Then, as $\varepsilon \to 0^+$,*

$$\boldsymbol{H}_\varepsilon \overset{*}{\rightharpoonup} \boldsymbol{H} \qquad \text{weakly-} * \text{ in } L^\infty(0, T; \mathcal{H}_0^n), \tag{5.35a}$$

$$\boldsymbol{E}_\varepsilon \rightharpoonup \boldsymbol{E} \qquad \text{weakly in } L^2(0, T; \mathcal{H}_0^n), \tag{5.35b}$$

$$\frac{\partial \boldsymbol{H}_\varepsilon}{\partial t} \rightharpoonup \frac{\partial \boldsymbol{H}}{\partial t} \qquad \text{weakly in } L^2(0, T; \mathcal{H}_0^{n-1}), \tag{5.35c}$$

where $(\boldsymbol{E}, \boldsymbol{H})$ is the solution of (5.6) with boundary conditions (5.9) and initial conditions (5.7).

Proof Let $s = (s_1, s_2, s_3)$ be a multi-index of lentgth $|s| = s_1 + s_2 + s_3 \le n$. We use notation $\partial^s = \frac{\partial^{s_1}}{\partial x_1^{s_1}} \frac{\partial^{s_2}}{\partial x_1^{s_2}} \frac{\partial^{s_3}}{\partial x_1^{s_3}}$. Let $\delta > 0$, and let $\boldsymbol{J}_s = \partial^s \boldsymbol{J}_0 \star \rho_\delta$. Then $\boldsymbol{E}_s = \partial^s \boldsymbol{E}_\varepsilon \star \rho_\delta$ and $\boldsymbol{H}_s = \partial^s \boldsymbol{H}_\varepsilon \star \rho_\delta$ are weak solutions of the system

$$\begin{cases} \nabla \times \boldsymbol{H}_s - \varepsilon \partial_t \boldsymbol{E}_s - \sigma_0 \boldsymbol{E}_s = \boldsymbol{f}_s & \text{in } \Omega \times (0, T), \\ \mu \dfrac{\partial \boldsymbol{H}_s}{\partial t} + \nabla \times \boldsymbol{E}_s = 0, & \text{in } \Omega \times (0, T), \\ \boldsymbol{n} \times \boldsymbol{H}_s = 0, & \text{on } \partial\Omega \times (0, T), \end{cases} \tag{5.36}$$

with the source being defined by $f_s = J_s - (\sigma - \sigma_0)\psi_s - R_s$, where

$$R_s = \sum_{\substack{\beta \leq s \\ \beta \neq s}} \binom{s}{\beta} (\partial^{s-\beta}\sigma\partial^{\beta}\psi) \star \rho_\delta.$$

As a consequence of the energy identity

$$\frac{d}{dt}\left(\varepsilon\|E_s\|_{L^2}^2 + \|H_s\|_{L^2}^2\right) + 2\,(\sigma E_s,\,E_s)_{L^2} = 2\,(J_s + R_s,\,E_s)_{L^2}$$

after a finite recursion argument we arrive at

$$\sum_{|s|\leq n}\left[\frac{d}{dt}\left(\varepsilon\|E_s\|_{L^2}^2 + \|H_s\|_{L^2}^2\right) + \|E_s\|_{L^2}^2\right] \leq c_1\|J_s\|_{\mathscr{H}_0^n}^2$$

where c_1 depends on σ_0 and on n, only. We recall that for every $\psi \in \mathscr{H}_0^n$

$$c_2^{-1}\|\psi\|_{\mathscr{H}_0^n}^2 \leq \sum_{|\alpha|\leq n}\|\partial^s\psi\|_{L^2}^2 \leq c_2\|\psi\|_{\mathscr{H}_0^n}^2$$

where c_2 only depends on n. Hence, in view of (5.33),

$$\sum_{|s|\leq n}\left[\sup_{0\leq t\leq T}\left(\varepsilon\|E_s\|_{L^2}^2 + \|H_s\|_{L^2}^2\right) + \int_0^T\|E_s\|_{L^2}^2\,dt\right] \leq 2c_1\int_0^T\|J\|_{\mathscr{H}_0^n}^2\,dt + 2c_2A_0^2.$$

where c_2 is an absolute constant. Thus, by setting

$$c_3 = \sqrt{2c_2\left(c_1\int_0^T\|J\|_{\mathscr{H}_0^n}^2\,dt + A_0^2\right)},$$

the mapping that takes every pair of fields (φ, ψ), with $\varphi(0) = E_\varepsilon$ and $\psi(0) = H_\varepsilon$, to the solution of system (5.4), with (5.9), subject to the initial conditions (5.5), maps the space

$$\left\{(\varphi, \psi) \in \mathcal{X} : \|\sqrt{\varepsilon}\varphi\|_{L^\infty(0,T;\mathscr{H}_0^n)} + \|\psi\|_{L^2(0,T;\mathscr{H}_0^n)} + c_1\|\varphi\|_{L^2(0,T;\mathscr{H}_0^n)} \leq c_3\right\}, \tag{5.37}$$

where $\mathcal{X} = C([0, T]; \mathscr{H}_0^n \times \mathscr{H}_0^n) \cap C^1([0, T]; \mathscr{H}_0^{n-1} \times \mathscr{H}_0^{n-1})$, into itself. Also, it is easily seen that this mapping is a contraction if we equip the vector space (5.37) with the distance induced by the norm

$$(\varphi, \psi) \longmapsto \|(\sqrt{\varepsilon}\varphi,\,\psi)\|_{L^\infty(0,T;L^2(\Omega;\mathbb{R}^3\times\mathbb{R}^3))}.$$

By uniqueness, we infer the estimate

$$\limsup_{\varepsilon \to 0} \left[\sup_{0 \le t \le T} \left(\varepsilon \|E_\varepsilon\|^2_{\mathscr{H}^n_0} + \|H_\varepsilon\|^2_{\mathscr{H}^n_0} \right) + \int_0^T \|E_\varepsilon\|^2_{\mathscr{H}^n_0} \, dt \right]$$

$$\le 2c_1 c_2 \int_0^T \|J\|^2_{\mathscr{H}^n_0} \, dt + 2c_2 \cdot A_0^2.$$

By Banach-Alaouglu theorem, we deduce that (5.35) holds, along a suitable sequence $\varepsilon_j \to 0^+$, for an appropriate limit (E, H). The fact that the limit solves (5.6) with (5.9) is a consequence. By uniqueness, (5.35) holds then for any sequence $\varepsilon_j \to 0^+$, as desired. □

Remark 5.4.2 As a consequence of the proof of Theorem 5.4.1 and of the second equation in (5.4a), there exists a constant $c > 0$, depending only σ_0, such that the solution satisfies the estimate

$$\sup_{0 \le t \le T} \left(\varepsilon \|E_\varepsilon\|^2_{\mathscr{H}^3_0} + \|H_\varepsilon\|^2_{\mathscr{H}^3_0} \right) + \int_0^T \|E_\varepsilon\|^2_{\mathscr{H}^3_0} \, dt + \int_0^T \left\| \frac{\partial H_\varepsilon}{\partial t} \right\|^2_{\mathscr{H}^2_0} \, dt \le c^2$$

(5.38)

for all $\varepsilon \in (0, c^{-1})$, provided that

$$\int_0^T \|J\|^2_{\mathscr{H}^n_0} \, dt + A_0^2 \le c.$$

The material above suggests one to consider the formal expansion of solutions of the hyperbolic Maxwell system (5.4)

$$E_\varepsilon = E + \sqrt{\varepsilon}\varphi + o(\sqrt{\varepsilon}), \qquad H_\varepsilon = H + \sqrt{\varepsilon}\psi + o(\sqrt{\varepsilon}),$$

in which the solution of the magneto-quasistatic Maxwell system (5.6) is the first term. We see that the higher order term (φ, ψ) is provided, formally, by equations

$$\begin{cases} \nabla \times \psi - \sigma\varphi = \sqrt{\varepsilon}\partial_t E, \\ \nabla \times \varphi + \mu\dfrac{\partial \psi}{\partial t} = 0, \\ n \times \psi = 0. \end{cases}$$

(5.39)

In particular, by applying the parabolic estimate (5.24) of Theorem 5.3.3 to the system (5.39), and by combining the result with (5.38), we see that for the hope of giving rigour to the expansion to be legit it becomes relevant to consider the limit

$$\lim_{\varepsilon \to 0^+} \frac{H_{0\varepsilon} - H_0}{\sqrt{\varepsilon}}.$$

The existence of the limit, with respect to an appropriate topology, may be useful in order to give an initial value to ψ and solve (5.39). This is related to the following theorem.

Theorem 5.4.3 (Singular Convergence)   *Under the assumptions made in Theorem 5.4.1, and under the additional assumption that*

$$\| \boldsymbol{H}_{0\varepsilon} - \boldsymbol{H}_0 \|_{\mathscr{H}_0^1} = O(\sqrt{\varepsilon}), \qquad as \; \varepsilon \to 0^+, \tag{5.40}$$

as $\varepsilon \to 0^+$ we have

$$\| \boldsymbol{H}_\varepsilon - \boldsymbol{H} \|_{L^\infty(0,T;\mathscr{H}_0^1)} = O(\sqrt{\varepsilon}) \quad and \quad \| \boldsymbol{E}_\varepsilon - \boldsymbol{E} \|_{L^2(0,T;\mathscr{H}_0^1)} = O(\sqrt{\varepsilon}).$$

Moreover, for every $T_0 \in (0, T)$

$$\sup_{t\in[T_0,T]} \| \boldsymbol{E}_\varepsilon(t) - \boldsymbol{E}(t) \|_{\mathscr{H}_0^1} = O(\varepsilon^{1/4}), \qquad as \; \varepsilon \to 0^+.$$

Proof In view of Remark 5.4.2, after the limit procedure we know that

$$\boldsymbol{E} \in C(0, T; \mathscr{H}_0^2) \cap W^{1,2}(0, T; \mathscr{H}_0^1), \tag{5.41}$$

and that

$$\limsup_{\varepsilon\to0^+} \sup_{t\in[0,T]} \| \boldsymbol{E}_\varepsilon(t) \|_{\mathscr{H}_0^2} \le c_1, \qquad \int_0^T \| \partial_t \boldsymbol{E}(t) \|_{\mathscr{H}_0^1}^2 \, dt \le c_2. \tag{5.42}$$

for appropriate positive constants c_1 and c_2. Setting $\boldsymbol{\xi}_\varepsilon = \boldsymbol{E}_\varepsilon - \boldsymbol{E}$ and $\boldsymbol{\chi}_\varepsilon = \boldsymbol{H}_\varepsilon - \boldsymbol{H}$, we see that $\boldsymbol{\eta}_\varepsilon = \nabla \times \boldsymbol{\xi}_\varepsilon$ and $\boldsymbol{\zeta}_\varepsilon = \nabla \times \boldsymbol{\chi}_\varepsilon$ solve the system

$$\begin{cases} \nabla \times \boldsymbol{\zeta}_\varepsilon - \sigma \boldsymbol{\eta}_\varepsilon - \varepsilon \dfrac{\partial \boldsymbol{\eta}_\varepsilon}{\partial t} = \varepsilon \nabla \times \dfrac{\partial \boldsymbol{E}}{\partial t}, & \text{in } \Omega \times (0, T), \\[2mm] \mu \dfrac{\partial \boldsymbol{\zeta}_\varepsilon}{\partial t} + \nabla \times \boldsymbol{\eta}_\varepsilon = 0 & \text{in } \Omega \times (0, T), \\[2mm] \boldsymbol{n} \times \boldsymbol{\zeta}_\varepsilon = 0 & \text{on } \partial\Omega \times (0, T), \end{cases}$$

whence if follows that

$$\frac{1}{2}\frac{d}{dt}\left(\varepsilon \| \boldsymbol{\eta}_\varepsilon \|_{L^2}^2 + \mu \| \boldsymbol{\zeta}_\varepsilon \|_{L^2}^2 \right) + (\sigma \boldsymbol{\eta}_\varepsilon, \boldsymbol{\eta}_\varepsilon)_{L^2} = -\varepsilon \left(\nabla \times \frac{\partial \boldsymbol{E}}{\partial t}, \boldsymbol{\eta}_\varepsilon \right). \tag{5.43}$$

Then, thanks to (5.42), from this energy identity we deduce that

$$\sup_{t\in[0,T]} \left(\varepsilon \| \boldsymbol{\xi}_\varepsilon \|_{\mathscr{H}_0^1}^2 + \mu \| \boldsymbol{\chi}_\varepsilon \|_{\mathscr{H}_0^1}^2 \right) + \sigma_0 \int_0^T \| \boldsymbol{\xi}_\varepsilon \|_{\mathscr{H}_0^1} \, dt \le c_2 \cdot \varepsilon + \varepsilon \| \boldsymbol{\xi}_\varepsilon(0) \|_{\mathscr{H}_0^1}^2 + \| \boldsymbol{\chi}_\varepsilon(0) \|_{\mathscr{H}_0^1}^2.$$

Using now (5.40) and recalling the first inequality in (5.42), we infer

$$\sup_{t\in[0,T]} \left(\varepsilon \|\boldsymbol{\xi}_\varepsilon\|^2_{\mathcal{H}^1_0} + \|\boldsymbol{\chi}_\varepsilon\|^2_{\mathcal{H}^1_0} \right) + \int_0^T \|\boldsymbol{\xi}_\varepsilon\|_{\mathcal{H}^1_0} \, dt \le c_4 \cdot (c_1 + c_2 + c_3) \cdot \varepsilon$$

where the constant $c_4 > 0$ is independent of ε, which ends the first part of the proof.

In order to prove the last statement, we use (5.41) and (5.42) to estimate the right hand side in identity (5.43). By doing so, we find a constant $c_5 > 0$, independent of ε, such that

$$\varepsilon e^{-\frac{t}{\varepsilon}} \frac{d}{dt} \left(e^{\frac{t}{\varepsilon}} \|\boldsymbol{\xi}_\varepsilon(t)\|^2_{\mathcal{H}^1_0} \right) + \frac{d}{dt} \|\boldsymbol{\chi}_\varepsilon(t)\|^2_{\mathcal{H}^1_0} = \frac{d}{dt} \left(\varepsilon \|\boldsymbol{\xi}_\varepsilon\|^2_{\mathcal{H}^1_0} + \|\boldsymbol{\chi}_\varepsilon\|^2_{\mathcal{H}^1_0} \right) + \|\boldsymbol{\xi}_\varepsilon\|^2_{\mathcal{H}^1_0} \le c_5 \varepsilon \|\partial_t \boldsymbol{E}\|_{\mathcal{H}^1_0}.$$

Integrating in time, by Theorem 5.4.3 we arrive at

$$e^{t/\varepsilon} \|\boldsymbol{\xi}_\varepsilon(t)\|^2_{\mathcal{H}^1_0} \le \|\boldsymbol{\xi}_\varepsilon(0)\|^2_{\mathcal{H}^1_0} + c_5 \int_0^t e^{\tau/\varepsilon} \|\partial_t \boldsymbol{E}(\tau)\|_{\mathcal{H}^1_0} \, d\tau \, .$$

By (5.42) and Cauchy-Schwartz inequality, the latter implies

$$e^{t/\varepsilon} \|\boldsymbol{\xi}_\varepsilon(t)\|^2_{\mathcal{H}^1_0} \le \|\boldsymbol{\xi}_\varepsilon(0)\|^2_{\mathcal{H}^1_0} + c_5 \sqrt{\tfrac{c_2}{2}} \cdot e^{t/\varepsilon} \sqrt{\varepsilon} \qquad (5.44)$$

whence the conclusion.   □

Theorem 5.4.4 (Non-singular Convergence) *Under the assumptions of Theorem 5.4.3, and assuming furthermore that*

$$\|\nabla \times \boldsymbol{H}_{0\varepsilon} - \sigma \boldsymbol{E}_{0\varepsilon} - \boldsymbol{J}_0(0)\|_{\mathcal{H}^1_0} = O(\sqrt{\varepsilon}), \qquad as \ \varepsilon \to 0^+, \qquad (5.45)$$

we have

$$\sup_{t\in[0,T]} \|\boldsymbol{E}_\varepsilon(t) - \boldsymbol{E}(t)\|_{L^2(\Omega;\mathbb{R}^2)} = O(\sqrt{\varepsilon}), \qquad as \ \varepsilon \to 0^+.$$

Proof By triangle inequality, (5.40) and (5.45) imply

$$\|\sigma \boldsymbol{E}(0) - \sigma \boldsymbol{E}_{0\varepsilon}\|_{\mathcal{H}^1_0} = O(\varepsilon^{1/2}), \qquad as \ \varepsilon \to 0^+.$$

Then, repeating the proof of Theorem 5.4.3 all the way up to (5.44), we finally obtain

$$\|\boldsymbol{E}_\varepsilon(t) - \boldsymbol{E}(t)\|^2_{\mathcal{H}^1_0} \le C_1 e^{-t/\varepsilon} \varepsilon + C_2 \varepsilon^{1/2},$$

for suitable constants $C_1, C_2 > 0$.   □

5.5   A Forward Model with Singular Sources

We introduce in this section a model that suites the case of sources very concentrated in limited regions, rather than being spread over a wide area. To this aim, we consider sources described by vector-valued (current) distributions $J_0 \in \mathcal{E}'(\mathbb{R}^3 \times \mathbb{R})^3$ with compact support in $\Omega \times (0, T)$. In this case, solutions to the magneto-quasistatic Maxwell equations (5.6) are understood in the sense of distributions, viz.

$$\int \int_\Omega H \cdot \nabla \times \nabla \times (\sigma^{-1} \varphi) \, dx \, dt - \int \int_\Omega \mu H \cdot \frac{\partial \varphi}{\partial t} \, dx \, dt = \langle J_0, \varphi \rangle, \quad (5.46)$$

for all $\varphi \in C_0^\infty(\Omega \times (0, T))$, where $\langle \cdot, \cdot \rangle$ denotes the distributional duality pairing.

5.5.1   Subsurface Sources

In particular, J_0 may be concentrated along, and tangent to, a given surface Σ. If in addition it belongs to the space defined in (5.10), then formally a vector field $H \in W^{1,2}(0, T; L^2(\Omega; \mathbb{R}^3))$ solving Eq. (5.8) is such that

$$\mu \frac{\partial H}{\partial t} + \nabla \times (\sigma^{-1} \nabla \times H) = \nabla \times (\sigma^{-1} J_0), \quad \text{in } \mathcal{D}'((\Omega \setminus \Sigma) \times (0, T)) \quad (5.47)$$

together with the transmission conditions

$$[\![n \times H]\!] = J_0 \quad \text{across } \Sigma. \quad (5.48)$$

To see that (5.48) holds, we introduce the electric field by setting $E = \sigma^{-1}(\nabla \times H - J_0)$. Then

$$\int_\Omega H \cdot \nabla \times \varphi \, dx - \int_\Omega \sigma E \cdot \varphi \, dx = \int_\Sigma J_0 \cdot \varphi \, d\Sigma \quad (5.49)$$

for any arbitrary test field $\varphi \in C_0^\infty(\Omega; \mathbb{R}^3)$. Considering a region $D \subset \Omega$, with $\overline{D} \subset \Omega$, that is split into un upper part D^+ and a lower one D^- by Σ, the validity of the previous equation for all field supported in $D \setminus \Sigma$ implies that (5.6a) and (5.6b) hold locally in D^+ and in D^-. Thus, we have

$$\nabla \times H - \sigma E = 0, \quad \text{in } (\Omega \setminus \Sigma) \times (0, T),$$
$$\mu \frac{\partial H}{\partial t} + \nabla \times E = 0, \quad \text{in } \Omega \times (0, T). \quad (5.50)$$

Therefore, if in (5.49) we choose $\boldsymbol{\varphi}$ with support intersecting Σ, then integrating by parts and using the first equation in (5.50) we get

$$\int_{\Sigma} \boldsymbol{\varphi} \cdot [\![\boldsymbol{H} \times \boldsymbol{n}]\!] \, d\Sigma = \int_{\Sigma} \boldsymbol{\varphi} \cdot \boldsymbol{J}_0 \, d\Sigma \,,$$

and that implies (5.48) because $\boldsymbol{\varphi}$ can have any arbitrary trace along Σ.

5.5.2 Signals in Free Homogeneous Space

Incidentally, we consider a fictitious model in which signals are free to propagate in a free space, without boundary conditions, filled with a homogeneous isotropic medium.

We fix $\kappa > 0$ and we recall that the function $\Gamma_\kappa \in C^\infty(\mathbb{R}^3 \times \mathbb{R})$ defined by

$$\Gamma_\kappa(\boldsymbol{x}, t) = \begin{cases} \left(\dfrac{\kappa}{4\pi t} \right)^{\frac{3}{2}} e^{-\frac{\kappa}{4t}(x^2 + y^2 + z^2)}, & \text{if } \boldsymbol{x} = (x, y, z) \in \mathbb{R}^3 \text{ and } t > 0, \\ 0 & \text{if } \boldsymbol{x} = (x, y, z) \in \mathbb{R}^3 \text{ and } t \leq 0, \end{cases} \tag{5.51}$$

solves

$$\kappa \frac{\partial}{\partial t} \Gamma - \nabla^2 \Gamma = 0, \qquad \text{in } \mathbb{R}^3 \times (0, +\infty),$$

where ∇^2 denotes the Laplace operator, with

$$\lim_{t \to 0^+} \int_{\mathbb{R}^3} \Gamma(x, t) \phi(x) \, dx = \phi(0) \,.$$

As a consequence, given constants $\mu > 0$ and $\overline{\sigma} > 0$, a particular solution of

$$\mu \cdot \overline{\sigma} \cdot \frac{\partial}{\partial t} \boldsymbol{H} - \nabla^2 \boldsymbol{H}_{\sigma_0} = \nabla \times \boldsymbol{J}_0, \qquad \text{in } \mathcal{D}'(\mathbb{R}^3 \times \mathbb{R}),$$

is provided us setting $\boldsymbol{H}_{\sigma_0} = \Gamma_\kappa \star (\nabla \times \boldsymbol{J}_0)$, where $\kappa = \mu \cdot \overline{\sigma}$ and the vector valued convolution in space-time is defined for all $u \in C^\infty(\mathbb{R}^3 \times \mathbb{R})$ and for all $\boldsymbol{T} \in \mathcal{E}'(\mathbb{R}^3 \times \mathbb{R})^3$ componentwise, according to the formula

$$u \star \boldsymbol{T}(\boldsymbol{x}, t) = \int_{\mathbb{R}} \int_{\mathbb{R}^3} u(\boldsymbol{x} - \boldsymbol{y}, t - s) \boldsymbol{T}(\boldsymbol{y}, s) \, d\boldsymbol{y} \, ds \,.$$

We make the assumption that the source is concentrated in a region where the electric conductivity introduced in (5.14a) is isotropic and homogeneous, say

$\sigma(x) = \sigma_0$ at all points x belonging to the support of $\boldsymbol{J}_0$. Then, by the material above the auxiliary system

$$\boldsymbol{\nabla} \times \boldsymbol{H}_\kappa - \overline{\sigma}\, \boldsymbol{E}_\kappa = \boldsymbol{J}_0 , \qquad \text{in } \mathcal{D}'(\mathbb{R}^3 \times \mathbb{R}),$$

$$\boldsymbol{\nabla} \times \boldsymbol{E}_\kappa + \mu\, \frac{\partial}{\partial t} \boldsymbol{H}_\kappa = 0 , \qquad \text{in } \mathcal{D}'(\mathbb{R}^3 \times \mathbb{R}),$$

is equipped with a solution $(\boldsymbol{E}_\kappa, \boldsymbol{H}_\kappa) \in C^\infty((\mathbb{R}^3 \times \mathbb{R}) \setminus \mathrm{supp}(\boldsymbol{J}_0))$ given by

$$\boldsymbol{H}_\kappa = \Gamma_\kappa \star (\boldsymbol{\nabla} \times \boldsymbol{J}_0) , \tag{5.52a}$$

$$\boldsymbol{E}_\kappa = \frac{1}{\overline{\sigma}} (\boldsymbol{\nabla} \times \boldsymbol{H}_\kappa - \boldsymbol{J}_0) . \tag{5.52b}$$

Eventually, since $\boldsymbol{\nabla} \cdot \boldsymbol{J}_0 = 0$ we have

$$\boldsymbol{\nabla} \times \boldsymbol{H}_\kappa = \Gamma_\kappa \star (\boldsymbol{\nabla} \times \boldsymbol{\nabla} \times \boldsymbol{J}_0) = \Gamma_\kappa \star (-\nabla^2 \boldsymbol{J}_0) = -(\nabla^2 \Gamma_\kappa) \star \boldsymbol{J}_0 . \tag{5.53}$$

5.5.3 Renormalised Signals

Now we consider the signals propagating, virtually, in the medium defined by difference between the real medium and the fictitious one that was introduced in the previous section.

Given a solution of (5.6) in the sense of distributions, the fields defined by difference setting $\boldsymbol{u} = \boldsymbol{H} - \boldsymbol{H}_\kappa$ and $\boldsymbol{v} = \boldsymbol{E} - \boldsymbol{E}_\kappa$ solve a system of the form

$$\boldsymbol{\nabla} \times \boldsymbol{u} - \sigma \boldsymbol{v} = \boldsymbol{f} ,$$

$$\boldsymbol{\nabla} \times \boldsymbol{v} + \mu\, \frac{\partial}{\partial t} \boldsymbol{u} = 0 , \tag{5.54}$$

in $\mathcal{D}'(\Omega \times (0, T))$. Equivalently, the magnetic part $\boldsymbol{u}$ is dictated by the magneto-quasistatic problem

$$\mu\, \frac{\partial \boldsymbol{u}}{\partial t} - \boldsymbol{\nabla} \times (\sigma^{-1} \boldsymbol{\nabla} \times \boldsymbol{u}) = \boldsymbol{f} \tag{5.55}$$

In both (5.54) and (5.55), we set

$$\boldsymbol{f}(x) = \begin{cases} \mu(\overline{\sigma} - \sigma) \dfrac{\partial \Gamma_\kappa}{\partial t} \star \boldsymbol{J}_0 & \text{in } \{\sigma \not\equiv \sigma_0\} \\ 0 & \text{in } \{\sigma \equiv \sigma_0\} \end{cases} \tag{5.56}$$

By [3, Appendix A.1], the boundary values $u \times n = -H_\kappa \times n$ belong to the space $H^{-\frac{1}{2}}(\text{div}_\tau; \partial\Omega)$, defined in as in (5.10) with $\Sigma = \partial\Omega$, because their tangential divergence equals $-\nabla \cdot H_\kappa \times n$ and $\nabla \cdot H_\kappa = \Gamma_\kappa \star \nabla \cdot (\nabla \times J_0) = 0$. Hence they extend to a suitable function $F_\kappa \in L^2(0, T; \mathcal{H}_0^1)$. Setting $w = u - F_\kappa$ we then have

$$\nabla \times w - \sigma v = f + \nabla \times F_\kappa, \qquad \text{in } \Omega \times (0, T),$$

$$\nabla \times v + \mu \frac{\partial w}{\partial t} = 0, \qquad \text{in } \Omega \times (0, T), \tag{5.57}$$

$$w \times n = 0, \qquad \text{on } \partial\Omega \times (0, T).$$

Equivalently, we have that $w \in L^2(0, T; \mathcal{H}_0^1)$ and

$$\mu \frac{\partial w}{\partial t} + \nabla \times (\sigma^{-1} \nabla \times w) = \nabla \times (\sigma^{-1}(f + \nabla \times F_\kappa)). \tag{5.58}$$

Since we are assuming the singular source J_0 to be supported in the level set $\{\sigma(x) = \overline{\sigma}\}$, (5.56) implies $f \in L^\infty(\Omega \times (0, T))$, whence it follows that $f \in L^2(0, T; L^2(\Omega; \mathbb{R}^3))$, because Ω is bounded. Hence, by Theorem 5.3.3, there exists a unique $w \in W^{1,2}(0, T; \mathcal{H}_0^1)$ solving (5.58), with $w(0) = 0$. Then, setting $v = \sigma^{-1}(\nabla \times w - f + \nabla \times F_\kappa)$, we have

$$\int_\Omega w \cdot \nabla \times \varphi \, dx - \int_\Omega \sigma v \cdot \varphi \, dx = \int_\Omega (f + \nabla \times F_\kappa) \cdot \varphi \, dx$$
$$\int_\Omega v \cdot \nabla \times \psi \, dx + \int_\Omega \mu \frac{\partial}{\partial t} w \cdot \psi \, dx = 0, \tag{5.59}$$

for all $\varphi \in \mathcal{H}^1$, for all $\psi \in \mathcal{H}_0^1$, and for a.e. $t \in (0, T)$. We thus have proved the following.

Theorem 5.5.1 *Let $\Omega \subset \mathbb{R}^3$ satisfy (5.13), let $\mu, \overline{\sigma}$ be positive constants and set $\kappa = \mu \cdot \overline{\sigma}$. Let σ be a piecewise constant positive bounded function, and let $H_0 \in \mathcal{H}_0^1$. Let $J \in \mathcal{E}'(\mathbb{R}^3 \times \mathbb{R})$ be a vector-valued distribution with support contained in the level set $\{\sigma(x) = \overline{\sigma}\}$. Then there exist a unique solution (E, H) of (5.6) in $\mathcal{D}'(\Omega \times (0, T))$, that takes the form $E = E_\kappa + v$, and*

$$H = H_\kappa + F_\kappa + w, \tag{5.60}$$

where

(i) (E_κ, H_κ) is defined as in (5.52),
(ii) F_κ is any element of $L^2(0, T; \mathcal{H}_0^1)$ agreeing with $-H_\kappa$ on $\partial\Omega \times (0, T)$, and
(iii) w is the solution of (5.58), with f being defined by (5.56), under the initial condition $w(0) = 0$.

Theorem 5.5.1 provides a split formula (5.60) that can be used to compute the values of the magnetic field H due to the singular source J_0. The first two summands in the right hand side are the result of representation formulas involving convolution with explicit kernels, see (5.52a). The extra term w is the solution of the magneto-quasistatic system (5.59), with a diffuse source (also defined by convolution, see (5.56)), that falls in the theory discussed in Sect. 5.3.

We now report some basic strategy for the methods in numerical approximation of w. The following material does not make, by any means, any pretence of completeness. For a modern and comprehensive exposition of the topic, we refer the interested reader to the book [3]. For an introduction to the more general magneto-hydrodynamic setting we mention the nice treatise [18].

5.6 Inverse Source Problems

5.6.1 Nonuniqueness of Volume Currents

It is known since the work of Helmholtz that the reconstruction of electric sources from tangential boundary measurements is generally an ill-posed problem. For the existence of non-radiating volume source currents in time-harmonic regime, we refer to [1] in the complete hyperbolic setting and to [4] in the eddy current case. There are little differences between the problem considered in the literature and the one surveyed in these pages, but we present however some details on this topic. An expedient integral identity holds: denoting by Θ^* the adjoint mapping to the linear "Dirichlet-to-Neumann" map

$$n \times H \longmapsto n \times \left(\sigma^{-1}\nabla \times H\right)$$

we see that the forward parabolic equation (5.8) with initial condition $H(0) = 0$, implies the identity

$$\int_0^T \int_\Omega H \cdot \mathcal{P}^*(\boldsymbol{\xi})\,dx\,dt + \int_0^T \left\langle \sigma^{-1}\nabla \times \boldsymbol{\xi} + \Theta^*(\boldsymbol{\xi})\,,\,n \times H\right\rangle dt = \int_0^T \int_\Omega \nabla \times (\sigma^{-1}J_0) \cdot \boldsymbol{\xi}\,dx\,dt\,, \tag{5.61}$$

for all smooth vector fields $\boldsymbol{\xi}$ such that $\boldsymbol{\xi}(T) = 0$. We introduced the backward parabolic operator

$$\mathcal{P}^*(\boldsymbol{\xi}) = -\mu\partial_t\boldsymbol{\xi} + \nabla \times \left(\sigma^{-1}\nabla \times \boldsymbol{\xi}\right). \tag{5.62}$$

Thus, tangential boundary measurements will be blind to any source that is orthogonal, in a suitable fractional Sobolev space, to the kernel of the adjoint operator $\mathcal{P}^*$ with vanishing final data

$$\mathcal{K}(\Omega) = \left\{\boldsymbol{\xi} \in W^{1,2}(0, T; \mathscr{H}_0^1) : \mathcal{P}^*(\boldsymbol{\xi}) = 0\,,\ \boldsymbol{\xi}(T) = 0\right\}.$$

For a more precise appreciation of this point, it is convenient to introduce the space $\mathscr{S}(\Omega)$ of *radiating sources*, consisting of the closure in $L^2(0, T; L^2(\Omega; \mathbb{R}^3))$ of the vector space

$$\left\{ \boldsymbol{\xi} \in \mathcal{K}(\Omega) : \boldsymbol{\xi} = \nabla \times (\sigma^{-1} \boldsymbol{v}) \text{ for some } \boldsymbol{v} \in L^2(0, T; L^2(\Omega; \mathbb{R}^3)) \right\}.$$

It can be seen that

$$\mathscr{N}(\Omega) = \{ \boldsymbol{g} \in L^2(0, T; L^2(\Omega; \mathbb{R}^3)) : (\boldsymbol{g}, \boldsymbol{f})_{L^2} = 0, \text{ for all } \boldsymbol{f} \in \mathscr{S}(\Omega) \}$$

is not a trivial space. $\mathscr{N}(\Omega)$ is said to collect *non-radiating sources* because of the following result.

Theorem 5.6.1 *Let $\Omega \subset \mathbb{R}^3$ satisfy (5.13), let $\mu > 0$, let $\sigma \in L^\infty(\Omega; \mathbb{R}^{3\times3})$, and assume that (5.14) holds. Let $\boldsymbol{J}_0 \in L^2(0, T; L^2(\Omega; \mathbb{R}^3))$, with $\nabla \times (\sigma^{-1}\boldsymbol{J}_0) = \boldsymbol{f} + \boldsymbol{g}$ where $\boldsymbol{f} \in \mathscr{S}(\Omega)$ and $\boldsymbol{g} \in \mathscr{N}(\Omega)$, and let $(\boldsymbol{E}, \boldsymbol{H})$ be the solution of (5.6), with (5.9), subject to the initial condition $\boldsymbol{H}(0) = 0$. Then*

(i) the knowledge of $\boldsymbol{n} \times \boldsymbol{H}$ on $\partial\Omega \times (0, T)$ uniquely determines $\boldsymbol{f}$;
(ii) if $\boldsymbol{f} = 0$, then $\boldsymbol{n} \times \boldsymbol{H} = 0$ on $\partial\Omega \times (0, T)$.

Proof If $\boldsymbol{n} \times \boldsymbol{H} = 0$ on $\partial\Omega \times (0, T)$, then by (5.61) we have $(\boldsymbol{f}, \boldsymbol{\xi})_{L^2} = 0$ for all $\boldsymbol{\xi} \in \mathscr{S}(\Omega)$ and that implies (i). As for (ii), we note that, for every element $\boldsymbol{\eta}$ of the trace space defined by (5.10) with $\Sigma = \partial\Omega$, the backward-in-time parabolic problem

$$\begin{cases} \mathcal{P}^*(\boldsymbol{\xi}) = 0 & \text{in } \Omega \times (0, T), \\ \boldsymbol{\xi}(T) = 0 & \text{in } \Omega \times \{T\}, \\ \sigma^{-1}\nabla \times \boldsymbol{\xi} + \Theta^*(\boldsymbol{\xi}) = \boldsymbol{\eta} & \text{on } \partial\Omega \times (0, T), \end{cases}$$

has a unique solution, for which (5.61) implies

$$\int_0^T \langle \boldsymbol{\eta}, \boldsymbol{n} \times \boldsymbol{H} \rangle = 0.$$

Since $\boldsymbol{\eta}$ was arbitrary, we obtain that $\boldsymbol{n} \times \boldsymbol{H} = 0$. □

5.6.2 Uniqueness for Atomic Sources

In view of non-uniqueness of volume currents, when considering the inverse problem one is lead to postulate a priori assumptions on the structure of the source.

In particular, we restrict our attention to sources that satisfy precise properties related to

- the geometry of their support;
- the vector-valued information on their orientation
- the scalar information about their intensity as currents.

The simplest situation is related to a source with a trivial orientation, which brings one to consider scalar equations, rather than systems. Also, we may assume a uniform electric current density. Then, we are led to the consider the following equation

$$\mu_0\sigma_0 \frac{\partial u}{\partial t}(\boldsymbol{x}, t) - \nabla^2 u(\boldsymbol{x}, t) = 0, \qquad (\boldsymbol{x}, t) \in \mathbb{R}^3 \times (0, T)$$

$$u(\boldsymbol{x}, 0) = \sum_{j=1}^{N} \delta(\boldsymbol{x} - \boldsymbol{a}_j), \qquad \text{in } \mathbb{R}^3, \tag{5.63}$$

where the positive integer N and the points $\boldsymbol{a}_1, \ldots, \boldsymbol{a}_N$ are unknown, and δ is Dirac's delta.

TOY PROBLEM Given points $\boldsymbol{x}_1, \boldsymbol{x}_2, \boldsymbol{x}_3 \in \mathbb{R}^2 \times \{0\}$ in general position, and given a sequence $(t_j)_j \subset (0, T)$ **determine** N and $\boldsymbol{a}_1, \ldots, \boldsymbol{a}_N$ from the knowledge of $u(\boldsymbol{x}, t_1), u(\boldsymbol{x}, t_2), u(\boldsymbol{x}, t_3)$, etc. assuming that (5.63) holds.

The toy problem is easier to consider if we further assume that $N = 1$. Then, given $\boldsymbol{a}$ and $\boldsymbol{b}$ in the horizontal plane, the corresponding solutions of (5.63) are given by

$$u(\boldsymbol{x}, t) = \Gamma(\boldsymbol{x} - \boldsymbol{a}, \mu\sigma_0 t), \qquad v(\boldsymbol{x}, t) = \Gamma(\boldsymbol{x} - \boldsymbol{b}, \mu\sigma_0 t),$$

where Γ is as in (5.51) with $\kappa = \mu_0\sigma_0$. Then, by assumption, the three points $\boldsymbol{x}_1$, $\boldsymbol{x}_2$, and $\boldsymbol{x}_3$ are at the same euclidean distance r to $\boldsymbol{a}$: in particular, as in Fig. 5.1, the point $\boldsymbol{a}$ belongs to the vertical circle obtained intersecting the spheres of radius r centred at $\boldsymbol{x}_1$ and $\boldsymbol{x}_2$, respectively. If $\boldsymbol{x}_3$ is not on the line joining $\boldsymbol{x}_1$ and $\boldsymbol{x}_3$, then the sphere of radius centred at $\boldsymbol{x}_3$ hits the circle exactly at two points, among which the relevant one is $\boldsymbol{a}$.

If $N \geq 1$ is unknown, we assume that there exist two unordered sets of points $A = \{\boldsymbol{a}_1, \ldots, \boldsymbol{a}_N\}$ and $B = \{\boldsymbol{b}_1, \ldots, \boldsymbol{b}_M\}$ and we denote u, v the signals generated by A, B, respectively. The limit as $t \to \infty$ in the following identity between analytic functions

$$\sum_{j=1}^{N} e^{-\mu_0\sigma_0 \frac{|\boldsymbol{x}_1 - \boldsymbol{a}_j|^2}{4t}} = \sum_{j=1}^{M} e^{-\mu_0\sigma_0 \frac{|\boldsymbol{x}_1 - \boldsymbol{b}_j|^2}{4t}} \tag{5.64}$$

Fig. 5.1 The case $N = 1$

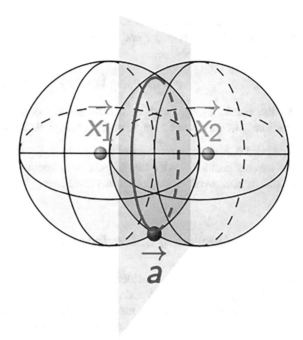

gives $N = M$. Before sending $t \to 0^+$, we may arrange the points in increasing distance to, say, the first reference point x_1

$$|a_1 - x_1| \leq |a_2 - x_1| \leq \ldots \leq |a_N - x_1|$$
$$|b_1 - x_1| \leq |b_2 - x_1| \leq \ldots \leq |b_N - x_1|$$

Denoting by n_1 (resp, m_1) the maximal integer n (resp. m) for which $|a_n - x_1| = |a_1 - x_1| =: R_1$ (resp., for which $|b_m - x_1| = |b_1 - x_1| =: S_1$), we have

$$\exp\left[-\mu_0\sigma_0\frac{R_1^2}{4t}\right](n_1 + o(1)) = \exp\left[-\mu_0\sigma_0\frac{S_1^2}{4t}\right](m_1 + o(1)), \qquad \text{as } t \to 0^+,$$

whence it follows that $R_1 = S_1$ and $n_1 = m_1$. Then from (5.64) we have arrived at

$$\sum_{j=n_1+1}^{N} e^{-\mu_0\sigma_0\frac{|x_1-a_j|^2}{4t}} = \sum_{j=n_1+1}^{N} e^{-\mu_0\sigma_0\frac{|x_1-b_j|^2}{4t}}. \qquad (5.65)$$

We can repeat this argument starting, this time, from (5.65) rather than from (5.64). By a finite descent, we conclude that there exist $k \leq N$, k positive integers $n_1 < \ldots < n_k$, and k positive numbers $R_1, \ldots, R_k$ such that all points a_j and b_j, with $n_\ell < j \leq n_{\ell+1}$, are at the same distance R_ℓ to x_1, for $\ell = 1, \ldots, k - 1$.

Also, we can arrive at similar conclusions arguing as done above except for replacing x_1 by either of x_2 and x_3. Therefore, for every single $j \in \{1, \ldots, N\}$ we have

$$|x_1 - a_j| = |x_2 - a_j| = |x_3 - a_j| \tag{5.66}$$

As the points x_1, x_2, and x_3 are in general position by assumption, as seen previously in the case of a single source (5.66) determines a_j. Thus, $A = B$.

Remark 5.6.2 For the purposes of applications, it would be interesting to consider the inverse problem under the "dipole assumption" in the case of non-constant conductivities: we do not consider this issue here.

5.6.3   Uniqueness of Subsurface Currents

It might be relevant to consider sources that are concentrated along surfaces such as fault planes, and to reconstruct them from boundary measurements. We mention a result that applies to media described by smooth constitutive coefficients. Uniqueness in this case results are available for surface currents that are a priori concentrated on the boundaries of subdomains: the following result, for example, requires the knowledge of a continuous set of measurements, that holds for sources of the "separable" form

$$J_0(x, t) = h(t) f(x). \tag{5.67}$$

Theorem 5.6.3 *Let Ω satisfy (5.13), and let μ and σ be smooth positive functions satisfying (5.14). Let $B \subset \Omega$ be a connected open set with a Lipschitz regular boundary Σ. Assume that $J_0 \in W^{1,2}(0, T; L^2(\Omega))$ is of the form (5.67) and that H is a solution of (5.6), with (5.9), subject to the initial conditions $H(0) = H_0$. Then, the knowledge of*

$$m(t, \psi) = \int_\Sigma \mu H \times v \cdot \psi \, d\Sigma \tag{5.68}$$

for all $t \in [0, T]$ and for all $\psi \in \mathcal{H}_0^1$, uniquely determines J_0.

Proof To replicate the proof done in [21] in hyperbolic setting, one needs the estimates (5.24), (5.25) for the magnetic field and the higher order estimate

$$\|H\|_{W^{1,\infty}(0,T;\mathcal{H}_0^1)} + \|H\|_{W^{1,2}(0,T;\mathcal{H}_0^2)} \le c(\psi) \left(\int_0^T h^2 \, dt \right)^{\frac{1}{2}}$$

that holds under the additional assumption that σ is smooth (for, see the proof of Theorem 5.4.1).   □

5.6.4 Inverse Source Problems with Controllability

A different class of inverse source problems presuppose the complete knowledge of the initial and final state of the system, and the full knowledge of the initial value of the source. For the following result, we refer to [22] (which focus on the electric field, with unessential changes).

Theorem 5.6.4 *Let Ω satisfy (5.13), let μ, σ be positive functions satisfying (5.14), let $H_0 \in L^2(\Omega)$, and let $J_0 \in W^{1,2}(0, T; L^2(\Omega))$ and let $H \in L^2(0, T; \mathcal{H}_0^1)$, with $\partial_t H \in L^2(0, T, (\mathcal{H}_0^1)^*)$, be a weak solution of (5.6) and (5.9), with (5.7). Then, there exists $T_0 \in (0, T)$ such that, for every $\tau < T_0$, the knowledge of the final state $H(\tau)$ and of the initial source $J_0(x, 0)$ determine uniquely J_0.*

Acknowledgments The uniqueness for atomic sources presented in Sect. 5.6.2 was illustrated to the author by Sergio Vessella in private communication. Some results commented in this paper were presented in Cetraro, Italy in Summer 2019, during the CIME-EMS Summer School in applied mathematics "Applied Mathematical Problems in Geophysics". The author is grateful to Massimo Chiappini, Roberto Carluccio, and Cesidio Bianchi from the Istituto Nazionale di Geofisica e Vulcanologia INGV for their interest in this research, and the Istituto Nazionale di Alta Matematica INDAM is acknowledged for the financial support.

References

1. R. Albanese, P.B. Monk, The inverse source problem for Maxwell's equations in magnetoencelography. SIAM J. Appl. Math. **62**, 1369–82 (2006)
2. G.S. Alberti, Hölder regularity for Maxwell's equations under minimal assumptions on the coefficients. Calc. Var. Partial Differ. Equ. **57**, 1–11 (2018)
3. A. Alonso Rodriguez, A. Valli, *Eddy Current Approximation of Maxwell Equations* (Springer-Verlag Italia, Milan, 2010)
4. A. Alonso Rodriguez, J. Camaño, A. Valli, Inverse source problems for eddy current equations. Inverse Probl. **28** (2012)
5. B. M. Brown, M. Marletta, J.M. Reyes, Uniqueness for an inverse problem in electromagnetism with partial data. J. Differ. Equ. **260**(8), 6525–6547 (2016)
6. M.A. Athanasiou, G.C. Anagnostopoulos, A.C. Iliopoulos, G.P. Pavlos, C.N. David, Enhanced ULF radiation observed by DEMETER two months around the strong 2010 Haiti earthquake. Nat. Hazards Earth Syst. Sci. **11**, 1091–1098 (2011)
7. W.H. Bakun et al., Implications for prediction and hazard assessment from the 2004 Parkfield earthquake. Nature **437**, 969–974 (2005)
8. M. Costabel, A coercive bilinear form for Maxwell's equations. J. Math. Anal. Appl. **157**, 527–541 (1991)
9. A.B. Draganov, U.S. Inan, N. Taranenko, Yu., ULF magnetic signatures at the Earth's surface due to ground waterflow. Geophys. Res. Lett. **18**, 1127–1130
10. G. Duvaut, J.L. Lions, *Inequalities in Mathematical Physics* (Springer-Verlag Berlin, Heidelberg, 1976)
11. L.C. Evans, *Partial Differential Equations*. AMS Graduate Studies in Mathematics, 2010
12. M.A. Fenoglio, M.J.S. Johnston, J.D. Byerlee, Magnetic and electric fields associated with changes in high pore pressure in fault zone - application to the Loma Prieta ULF emissions, in *Proc. of Workshop LXIII, Menlo Park, CA* 1994, pp. 262–278

13. A.C. Frasier-Smith, A. Bernardi, P.R. McGill, M. Ladd, R. Helliwell, O.G. Villard Jr., Low Frequency magnetic field measurement near the epicenter of the M_s 7.1 Loma Prieta earthquake. Geophys. Res. Lett. **17**(9), 1465–1468 (1990)
14. E. Francini, G. Franzina, S. Vessella, Existence and regularity for eddy currents system with non-smooth conductivity. SIAM J. Math. Anal. **52**(2), 2134–2157 (2020)
15. M.J.S. Johnston, R.J. Müller, Y. Sasai, Magnetic field observations in the near-field: the 28 June 1992 M_w 7.3 Landers ,California, Earthquake. Bull. Seism. Soc. Am. **8**, 792–798 (1994)
16. A. Milani, The quasi-stationary maxwell equations as singular limit of the complete equations: the quasi-linear case. J. Math. Anal. Appl. **102**, 251–274 (1984)
17. O.A. Molchanov, M. Hayakawa, Generation o fULF electromagnetic emission by microfracturing. Geophys. Rev Lett. **22**(22), 3091–3094 (1995)
18. C. Pagliantini, *Computational Magnetohydrodynamics with Discrete Differential Forms*. Ph.D. Thesis, ETH Zürich
19. P. Palangio, C. Di Lorenzo, F. Masci, M. Di Persio, The study of the electromagnetic anomalies linked with the Earth's crustal activity in the frequency band [0.001 Hz–100 kHz]. Nat. Hazards Earth Syst. Sci. **7**, 507–511 (2007)
20. G. Sebastiani, L. Malagnini, Forecasting the next Parkfield mainshock on the San Andreas fault (California). J. Ecol. Nat. Resour. **4**(6) (2020)
21. M. Slodička, M. Galba, Recovery of a time dependent source from a surface measurement in Maxwell's equations. Comp. Math. Appl. **71**, 368–380 (2016)
22. I.G. Stratis, A.N. Yannacopoulos, Some remarks on a class of inverse problems reelate to the parabolic approximation to the Maxwell equations: a controllability approach. Math. Methods Appl. Sci. **38**, 3866–3878 (2015)
23. R.H. Tatham, M.D. McCormack, Multicomponent Seismology in Petroleum Exploration. Society of Exploration Geophysicists, 1991
24. J.N. Thomas, J.J. Love, M.J.S. Johnston, On the reported magnetic precursor of the 1989 Loma Prieta earthquake. Phys. Earth Planet. Inter. **173**, 207–215 (2009)
25. N. Walker, V. Kadirkamanathan, O.A. Pokhotelov, Changes in the ultra-low frequency wave field during the precursorphase to the Sichuan earthquake: DEMETER observations. Ann. Geophys. **31**, 1597–1603 (2013)

Chapter 6
Conservation Laws in Continuum Mechanics

Giuseppe Maria Coclite and Francesco Maddalena

Abstract A general fundamental mathematical framework at the base of the conservation laws of continuum mechanics is introduced. The notions of weak solutions, and the issues related to the entropy criteria are discussed in detail. The spontaneous creation of singularities, and the occurrence of diffusive limits are explained in view of their physical implications. A particular emphasis is given to the applications of hyperbolic conservation laws in the models of gas dynamics, nonlinear elasticity and traffic flows.

6.1 Introduction

The conceptual structure informing continuum physics rests on two fundamental pillars: *balance laws* (or *conservation laws*) and *constitutive laws*. While the constitutive laws, ruling the specific properties of the material in which the physical phenomenon occurs (e.g. viscous fluids, elastic solids, elastic dielectric, etc.) are exposed to a great variety of possible relations (may be escaping any tentative of a definitive general theory), conservation laws admits a clear mathematical statement in the format of partial differential equations. In the general multidimensional spatial setting, an homogeneous hyperbolic conservation law takes the form [3, 19, 21]

$$\partial_t u + \sum_{\alpha=1}^{d} \partial_\alpha F_\alpha(u) = 0, \qquad (6.1)$$

G. M. Coclite (✉) · F. Maddalena
Department of Mechanics, Mathematics and Management, Polytechnic University of Bari, Bari, Italy
e-mail: giuseppemaria.coclite@poliba.it; francesco.maddalena@poliba.it

© The Author(s), under exclusive license to Springer Nature Switzerland AG 2022
M. Chiappini, V. Vespri (eds.), *Applied Mathematical Problems in Geophysics*,
C.I.M.E. Foundation Subseries 2308,
https://doi.org/10.1007/978-3-031-05321-4_6

where the state variable u, taking values in $\mathbb{R}^m$ depends on the spatial variables $(x_1, \ldots, x_d)$ and time t, $F_1, \ldots, F_d$ are smooth maps from $\mathbb{R}^m$ to $\mathbb{R}^m$, ∂_t denotes $\partial/\partial t$ and ∂_α denotes $\partial/\partial x_\alpha$.

In these notes we shall focus on the one-dimensional spatial case, governed by the first order partial differential equation

$$\partial_t u + \partial_x f(u) = 0, \tag{6.2}$$

where $f \in C^2(\mathbb{R}^N; \mathbb{R}^N)$, $u : [0, \infty) \times \mathbb{R} \to \mathbb{R}^N$, and $N \geq 1$. The function $u = u(t, x)$ is termed *conserved quantity*, $f = f(u)$ *flux*. If $N = 1$ we say that (6.2) is a *scalar conservation law*, if $N > 1$ we say that (6.2) is a *system of conservation laws* and it stays for

$$\begin{cases} \partial_t u_1 + \partial_x f_1(u_1, \ldots, u_N) = 0, \\ \ldots\ldots \\ \partial_t u_N + \partial_x f_N(u_1, \ldots, u_N) = 0, \end{cases}$$

where

$$u = u(t, x) = (u_1(t, x), \ldots, u_N(t, x)),$$

$$f = f(u) = (f_1(u_1, \ldots, u_N), \ldots, f_N(u_1, \ldots, u_N)).$$

In this section we try to answer the following questions:

(Q.1) Why do we use the terms *conservation law, conserved quantity,* and *flux* for (6.2), u, and f, respectively?

(Q.2) Which kind of physical phenomena is (6.2) able to describe?

(Q.3) Which are the mathematical features of the solutions of (6.2)?

Let us answer to **(Q.1)**. If u is a smooth solution of (6.2) and $a < b$ we have that (see Fig. 6.1)

$$\frac{d}{dt} \int_a^b u(t, x)dx = \int_a^b \partial_t u(t, x)dx$$

$$= -\int_a^b \partial_x f(u(t, x))dx = f(u(t, a)) - f(u(t, b))$$

$$= [\text{inflow at } x = a \text{ and time } t]$$

$$- [\text{outflow at } x = b \text{ and time } t].$$

Fig. 6.1 Flow trough the end points

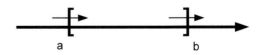

a   b

In other words, the conserved quantity u is neither created nor destroyed, the amount of u in the interval $[a, b]$ changes in function only of the flow through the two end points.

To answer to (**Q.2**) we proceed by showing some paradigmatic models founded in continuum mechanics, expressed in terms of conservation laws.

Rarefied Gas The simplest model of gasdynamic in one space dimension considers a material made of non interacting particles, idealizing a low dense gas. In the Lagrangian description, we can identify the particles using their initial position y. Let $\varphi(t, y)$ be the position at time t of the particle that at time $t = 0$ was in y, its velocity and acceleration are $\partial_t \varphi$ and $\partial^2_{tt} \varphi$, respectively. Since the particles do not interact within themselves, we cannot have two different particles in the same position at the same time, therefore $\varphi(t, \cdot)$ is increasing and, in particular, invertible. Let $\psi(t, \cdot)$ be the inverse of $\varphi(t, \cdot)$, i.e.,

$$y = \psi(t, \varphi(t, y))$$

and

$$x = \varphi(t, y) \Longleftrightarrow y = \psi(t, x).$$

Let $u(t, x)$ be the velocity of the particle at time t is in x, namely

$$x = \varphi(t, y),$$
$$u(t, x) = u(t, \varphi(t, y)) = \partial_t \varphi(t, y),$$
$$u(t, x) = \partial_t \varphi(t, \psi(t, x)).$$

The acceleration of the particle that at time t is in x is

$$\partial^2_{tt} \varphi(t, y) = \partial_t \Big(\partial_t \varphi(t, y) \Big) = \partial_t \Big(u(t, \varphi(t, y)) \Big)$$
$$= \partial_t u(t, \varphi(t, y)) + \partial_x u(t, \varphi(t, y)) \partial_t \varphi(t, y)$$
$$= \partial_t u(t, x) + \partial_x u(t, x) u(t, x).$$

Since the particles do not interact within themselves, there are no forces acting on them. Then, the *balance of linear momentum* delivers the equation

$$\partial_t u + \partial_x \left(\frac{u^2}{2} \right) = 0, \tag{6.3}$$

that is termed *Burgers equation* [5, 6, 18].

Traffic Flow 1 We begin with the road fluid-dynamic traffic model introduced by Lighthill, Whitham, and Richards [15, 17]. We consider a one way one lane infinite road. Let $\rho = \rho(t, x)$ be the the the density of vehicles at time t in the position x. Assuming that the vehicles behave as fluid particles we have [8, 9]

$$\partial_t \rho + \partial_x(\rho v) = 0, \tag{6.4}$$

where v is the velocity of the vehicles. The key assumption of Lighthill, Whitham, and Richards is that the velocity depends only on the density, namely

$$v = v(\rho), \tag{6.5}$$

that is somehow reasonable in case of highways. The drivers regulate their velocity in function of the number of vehicles in front of them. Therefore writing

$$f(\rho) = \rho v(\rho),$$

(6.4) reads

$$\partial_t \rho + \partial_x f(\rho) = 0. \tag{6.6}$$

On $v = v(\rho)$ it is reasonable to assume that

$$v(0) = v_{max}, \qquad v(\rho_{max}) = 0, \qquad v \text{ is decreasing.}$$

In particular, Lighthill, Whitham, and Richards proposed

$$v(\rho) = v_{max} \left(1 - \frac{\rho}{\rho_{max}} \right).$$

Compressible Non-viscous Gas The Lighthill-Whitham-Richards traffic model and the Burgers equation are model expressed in terms of scalar conservation laws, we continue by showing more models expressed in terms of systems of conservation laws.

The Euler equations for a non-viscous compressible gas in Lagrangian coordinates are

$$\begin{cases} \partial_t v - \partial_x u = 0, & \text{(conservation of mass)} \\ \partial_t u + \partial_x p = 0, & \text{(conservation of momentum)} \\ \partial_t \left(e + \frac{u^2}{2} \right) + \partial_x(up) = 0, & \text{(conservation of energy)} \end{cases} \tag{6.7}$$

where v is the specific volume (i.e., $1/v$ is the density), u is the velocity, e is the energy, and p is the pressure of the gas. Since we have three equations in four unknowns, we need a constitutive equation

$$p = p(e, v),$$

which selects the specific gas under consideration.

Nonlinear Elasticity Let us consider a one-dimensional elastic material body whose configuration in the Lagrangian description is represented by the displacement field $w(x, t)$. Then the strain measure is given by $u = \partial_x w$ and assuming the constitutive equation $\sigma = f(u)$ giving the Piola-Kirchhoff stress σ in terms of the strain measure u, the balance of linear momentum delivers the wave equation of motion [6, 10, 16]

$$\partial_{tt}^2 w - \partial_x f(u) = 0. \tag{6.8}$$

Setting $v = \partial_t w$ the velocity field, the previous wave equation takes the form of the following system of conservation laws

$$\begin{cases} \partial_t u - \partial_x v = 0, \\ \partial_t v - \partial_x f(u) = 0. \end{cases}$$

Shallow Water Equations Let $h(x, t)$ be the depth and $u(x, t)$ the mean velocity of a fluid moving in a rectangular channel of constant breadth and inclination α of the surface. Let also C_f be the friction coefficient affecting the friction force originating by the interaction of the fluid with the bed and g the gravity acceleration. The equations governing the motion of the fluid are given by

$$\begin{cases} \partial_t h + u\partial_x h + h\partial_x u = 0, \\ \partial_t u + u\partial_x u + g\cos\alpha\,\partial_x h = g\sin\alpha - C_f^2(u^2/h). \end{cases}$$

In the shallow water theory, the height of the water surface above the bottom is assumed to be small with respect to the typical wave lengths and the terms representing the slope and the friction are neglected giving rise to the simplified equations [20]

$$\begin{cases} \partial_t c + u\partial_x c + (c\partial_x u/2) = 0, \\ \partial_t u + u\partial_x u + 2c\partial_x c = 0, \end{cases}$$

where $c(x, t) = \sqrt{gh(x, t)}$.

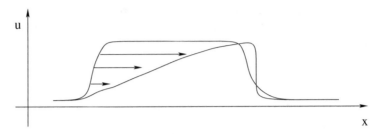

Fig. 6.2 Spontaneous creation of discontinuity in finite time

Traffic Flow 2 Finally, we have the traffic model proposed by Aw and Rascle [1]

$$
\begin{cases}
\partial_t \rho + \partial_x \left(y + \rho^{\gamma+1} \right) = 0, \\
\partial_t y + \partial_x \left(\dfrac{y^2}{2} - y\rho^{\gamma} \right) = 0,
\end{cases}
\tag{6.9}
$$

where ρ is the density, y this the generalized momentum of the vehicles, and γ is a positive constant.

Regarding (**Q.3**), one of the main features exhibited by hyperbolic of conservation laws is the possible creation of discontinuities. Indeed, even scalar problems with analytic flux and initial condition, like

$$
\begin{cases}
\partial_t + \partial_x \left(\dfrac{u^2}{2} \right) = 0, \quad t > 0, \ x \in \mathbb{R}, \\
u(0, x) = \dfrac{1}{1 + x^2}, \qquad x \in \mathbb{R},
\end{cases}
\tag{6.10}
$$

experience the creation of discontinuities in finite time [5, 6, 18], see Fig. 6.2.

The next sections are organized as follows. In Sect. 6.2 we introduce weak and entropy solutions and prove the classical uniqueness result of Kružkov. In Sect. 6.3 we introduce and solve the Riemann problem. In Sect. 6.4 we present one of the many different approaches to the existence issue: the vanishing viscosity. Finally, some elementary facts on BV functions are collected in the Appendix.

6.2 Entropy Solutions

We pointed out in Sect. 6.1 that even a Cauchy problem of the type

$$
\partial_t u + \partial_x \left(\dfrac{u^2}{2} \right) = 0, \qquad u(0, x) = \dfrac{1}{1 + x^2},
$$

with analytic flux ($u \mapsto u^2/2$) and analytic initial condition ($x \mapsto 1/(1+x^2)$) may experience discontinuities in finite time. It appears evident that additional physical and mathematical conditions must be required in order to reach a meaningful concept of solution. As a consequence we develop a wellposedness theory for conservation laws in the framework of *entropy solutions*, that are special distributional solutions satisfying suitable additional inequalities (or E-conditions). The definition is inspired by the Second Law of Thermodynamics, we consider only the distributional solutions along which the entropies decrease. Note that the physical entropies are all concave maps, in the mathematical community the entropies are assumed to be convex, this explain the discrepancy between the usual Second Law of Thermodynamics and the ones considered here.

6.2.1 Weak Solutions

Consider the scalar conservation law

$$\partial_t u + \partial_x f(u) = 0, \qquad t > 0, \ x \in \mathbb{R}, \tag{6.11}$$

endowed with the initial condition

$$u(0, x) = u_0(x), \qquad x \in \mathbb{R}, \tag{6.12}$$

and assume

$$f \in C^2(\mathbb{R}), \qquad u_0 \in L^\infty_{loc}(\mathbb{R}). \tag{6.13}$$

Definition 6.2.1 A function $u : [0, \infty) \times \mathbb{R} \to \mathbb{R}$ is a weak solution of the Cauchy problem (6.11) and (6.12), if

(i) $u \in L^\infty_{loc}((0, \infty) \times \mathbb{R})$;
(ii) u satisfies (6.11) and (6.12) in the sense of distributions in $[0, \infty) \times \mathbb{R}$, namely for every test function $\varphi \in C^\infty(\mathbb{R}^2)$ with compact support we have

$$\int_0^\infty \int_{\mathbb{R}} (u\partial_t\varphi + f(u)\partial_x\varphi) \, dt dx + \int_{\mathbb{R}} u_0(x)\varphi(0, x)dx = 0.$$

We say that u is a weak solution of the conservation law (6.11) if i) holds and

(iii) u satisfies (6.11) in the sense of distributions in $(0, \infty) \times \mathbb{R}$, namely for every test function $\varphi \in C^\infty((0, \infty) \times \mathbb{R})$ with compact support we have

$$\int_0^\infty \int_{\mathbb{R}} (u\partial_t\varphi + f(u)\partial_x\varphi) \, dt dx = 0.$$

Direct consequence of the Dominate Converge Theorem is the following.

Theorem 6.2.1 *Let $\{u_\varepsilon\}_{\varepsilon>0}$ and u be functions defined on $[0, \infty) \times \mathbb{R}$ with values in $\mathbb{R}$. If*

(i) there exists $M > 0$ such that $\|u_\varepsilon\|_{L^\infty((0,\infty)\times\mathbb{R})} \leq M$ for every $\varepsilon > 0$;
(ii) $u \in L^\infty((0, \infty) \times \mathbb{R})$;
(iii) $u_\varepsilon \to u$ in $L^1_{loc}((0, \infty) \times \mathbb{R})$ as $\varepsilon \to 0$;
(iv) every u_ε is a weak solution of (6.11);

then

$$u \text{ is a weak solution of } (6.11).$$

6.2.2   Rankine-Hugoniot Condition

The introduction of the notion of weak solution opens the possibility to deal with discontinuous functions which, as above remarked, naturally occur in the mathematics of conservation laws. Then in this section we analyze the *shocks*, that are the simplest discontinuous weak solutions of (6.11).

Let u_-, u_+, $\lambda \in \mathbb{R}$ be given and consider the function

$$U : [0, \infty) \times \mathbb{R} \longrightarrow \mathbb{R}, \qquad U(t, x) = \begin{cases} u_-, & \text{if } x < \lambda t, \\ u_+, & \text{if } x \geq \lambda t. \end{cases} \tag{6.14}$$

Since we are not interested to the trivial case $u_+ = u_-$ in the following we always assume

$$u_+ \neq u_-.$$

Theorem 6.2.2 (Rankine-Hugoniot Condition)   *The following statements are equivalent:*

(i) the function U defined in (6.14) is a weak solution of (6.11);
(ii) the following condition named Rankine-Hugoniot condition holds true, i.e.,

$$f(u_+) - f(u_-) = \lambda(u_+ - u_-). \tag{6.15}$$

Proof Let $\varphi \in C^\infty((0, \infty) \times \mathbb{R})$ be a test function with compact support. Consider the vector field

$$F = (U\varphi, f(U)\varphi)$$

and the domains

$$\Omega_+ = \{x > \lambda t\}, \qquad \Omega_- = \{x < \lambda t\}.$$

The definition of U gives

$$(t, x) \in \Omega_+ \implies \begin{cases} F(t, x) = (u_+\varphi, f(u_+)\varphi), \\ \text{div}_{(t,x)}(F)(t, x) = u_+\partial_t\varphi + f(u_+)\partial_x\varphi, \end{cases}$$

$$(t, x) \in \Omega_- \implies \begin{cases} F(t, x) = (u_-\varphi, f(u_-)\varphi), \\ \text{div}_{(t,x)}(F)(t, x) = u_-\partial_t\varphi + f(u_-)\partial_x\varphi. \end{cases}$$

Since

$$\partial\Omega_+ = \partial\Omega_- = \{x = \lambda t\},$$

and the outer normals to Ω_+ and Ω_- are $(\lambda, -1)$ and $(-\lambda, 1)$ we have

$$\int_0^\infty \int_\mathbb{R} (U\partial_t\varphi + f(U)\partial_x\varphi)dtdx$$

$$= \iint_{\Omega_+} (u_+\partial_t\varphi + f(u_+)\partial_x\varphi)dtdx + \iint_{\Omega_-} (u_-\partial_t\varphi + f(u_-)\partial_x\varphi)dtdx$$

$$= \iint_{\Omega_+} \text{div}(F)dtdx + \iint_{\Omega_-} \text{div}(F)dtdx$$

$$= \int_0^\infty (u_+, f(u_+)) \cdot (\lambda, -1)\varphi(t, \lambda t)dt + \int_0^\infty (u_-, f(u_-)) \cdot (-\lambda, 1)\varphi(t, \lambda t)dt$$

$$= [\lambda(u_+ - u_-) - (f(u_+) - f(u_-))] \int_0^\infty \varphi(t, \lambda t)dt.$$

Therefore

$$\int_0^\infty \int_\mathbb{R} (U\partial_t\varphi + f(U)\partial_x\varphi)dtdx = 0, \quad \forall\varphi$$

$$\Updownarrow$$

$$f(u_+) - f(u_-) = \lambda(u_+ - u_-),$$

that concludes the proof. $\square$

Remark 6.2.1 The Rankine-Hugoniot condition (6.15) is a scalar equation that links the right and left sates u_+, u_- and the speed λ of the shock. In particular, if f is Lipschitz continuous with Lipschitz constant L, (6.15) gives

$$|\lambda| = \frac{|f(u_+) - f(u_-)|}{|u_+ - u_-|} \leq L.$$

In other terms, the speed of propagation of the singularities is finite and varies between $-L$ and L.

Theorem 6.2.3 *Let* $u : [0, \infty) \times \mathbb{R} \to \mathbb{R}$, $\tau > 0$, $\xi \in \mathbb{R}$ *and* $U : [0, \infty) \times \mathbb{R} \longrightarrow \mathbb{R}$ *as defined in* (6.14). *If*

  (i) $u \in L^\infty_{loc}((0, \infty) \times \mathbb{R})$;
  (ii) u *is a weak solution of* (6.11);
 (iii) $\displaystyle \lim_{\varepsilon \to 0} \frac{1}{\varepsilon^2} \int_{-\varepsilon}^{\varepsilon} \int_{-\varepsilon}^{\varepsilon} |u(t + \tau, x + \xi) - U(t, x)| dt dx = 0$;

then (6.15) *holds.*

Proof For every $\mu > 0$ define

$$u_\mu(t, x) = u(\mu t + \tau, \mu x + \xi), \qquad t \geq -\frac{\tau}{\mu}, \ x \in \mathbb{R}.$$

Since u is a weak solution of (6.11), the same does u_μ. We claim that

$$u_\mu \longrightarrow U, \qquad f(u_\mu) \longrightarrow f(U), \qquad \text{in } L^1_{loc}((0, \infty) \times \mathbb{R}), \text{ as } \mu \to 0. \qquad (6.16)$$

Let $R > 0$ and $\mu < \frac{\tau}{R}$. Since

$$U(\mu t, \mu x) = U(t, x), \qquad t > 0, \ x \in \mathbb{R},$$

we get

$$\int_{-R}^{R} \int_{-R}^{R} |u_\mu(t, x) - U(t, x)| dt dx$$

$$= \frac{1}{\mu^2} \int_{-R\mu}^{R\mu} \int_{-R\mu}^{R\mu} |u(t + \tau, x + \xi) - U(t, x)| dt dx \longrightarrow 0,$$

namely

$$u_\mu \longrightarrow U, \qquad \text{in } L^1((-R, R) \times (-R, R)), \text{ as } \mu \to 0.$$

Therefore the Dominated Convergence Theorem gives (6.16). Finally, Theorem 6.2.1 and (6.16) implies that U is a weak solution of (6.11). Then, the claim follows from Theorem 6.2.2.   □

6.2.3   Nonuniqueness of Weak Solutions

In this section we show with a simple example that the Cauchy problem (6.11)–(6.12) may admit more than one weak solution.

Let us consider the Riemann problem for the Burgers equation

$$\partial_t u + \partial_x \left(\frac{u^2}{2} \right) = 0, \qquad u(0, x) = \begin{cases} 0, & \text{if } x < 0, \\ 1, & \text{if } x \geq 0. \end{cases} \qquad (6.17)$$

Thanks to Theorem 6.2.2 we know that the function

$$U(t, x) = \begin{cases} 0, & \text{if } x < t/2, \\ 1, & \text{if } x \geq t/2, \end{cases}$$

is a weak solution of (6.17).

Consider the function

$$v(t, x) = \begin{cases} 0, & \text{if } x < 0, \\ x/t, & \text{if } 0 \leq x < t, \\ 1, & \text{if } x \geq t. \end{cases}$$

Since for every test function $\varphi \in C^\infty(\mathbb{R}^2)$ with compact support

$$\int_0^\infty \int_{\mathbb{R}} \left(v \partial_t \varphi + \frac{v^2}{2} \partial_x \varphi \right) dt \, dx + \int_0^\infty \varphi(0, x) dx$$

$$= \int_0^\infty \left(\int_x^\infty \frac{x}{t} \partial_t \varphi \, dt \right) dx + \int_0^\infty \left(\int_0^t \frac{x^2}{2t^2} \partial_x \varphi \, dx \right) dt$$

$$+ \int_0^\infty \left(\int_0^x \partial_t \varphi \, dt \right) dx + \int_0^\infty \left(\int_t^\infty \partial_x \varphi \, dx \right) dt + \int_0^\infty \varphi(0, x) dx = 0,$$

then v is also a weak solution of (6.17).

6.2.4 Entropy Conditions

We showed in the previous section that the Cauchy problem (6.11)–(6.12) may admit more than one weak solution. In this section we introduce some additional conditions that will select the unique "physically meaningful" solution within the family of the weak solutions. Those conditions are inspired by the Second Law of Thermodynamics.

Definition 6.2.2 Let $\eta, q : \mathbb{R} \to \mathbb{R}$ be functions. We say that η is an entropy associated to (6.11) with flux q if

$$\eta, q \in C^2(\mathbb{R}), \qquad \eta'' \geq 0, \qquad \eta' f' = q'.$$

Remark 6.2.2 If u is a smooth solution of (6.11) and η is an entropy with flux q we have

$$\partial_t \eta(u) + \partial_x q(u) = 0.$$

Indeed

$$\partial_t \eta(u) + \partial_x q(u) = \eta'(u)\partial_t u + q'(u)\partial_x u$$
$$= \eta'(u)\left(\partial_t u + f'(u)\partial_x u\right)$$
$$= \eta'(u)\left(\partial_t u + \partial_x f(u)\right) = 0.$$

Definition 6.2.3 A function $u : [0, \infty) \times \mathbb{R} \to \mathbb{R}$ is an entropy solution of the Cauchy problem (6.11) and (6.12), if

(i) $u \in L^\infty_{loc}((0, \infty) \times \mathbb{R})$;

(ii) for every entropy η with flux q, u satisfies

$$\partial_t \eta(u) + \partial_x q(u) \leq 0, \qquad \eta(u(0, \cdot)) = \eta(u_0), \tag{6.18}$$

in the sense of distributions in $[0, \infty) \times \mathbb{R}$, namely for every nonnegative test function $\varphi \in C^\infty(\mathbb{R}^2)$ with compact support we have

$$\int_0^\infty \int_{\mathbb{R}} (\eta(u)\partial_t \varphi + q(u)\partial_x \varphi) \, dt dx + \int_{\mathbb{R}} \eta(u_0(x))\varphi(0, x)dx \geq 0. \tag{6.19}$$

We say that u is an entropy solution of the conservation law (6.11) if $i)$ holds and

(iii) for every entropy η with flux q, u satisfies

$$\partial_t \eta(u) + \partial_x q(u) \leq 0 \tag{6.20}$$

in the sense of distributions in $(0, \infty) \times \mathbb{R}$, namely for every nonnegative test function $\varphi \in C^\infty((0, \infty) \times \mathbb{R})$ with compact support we have

$$\int_0^\infty \int_\mathbb{R} (\eta(u)\partial_t \varphi + q(u)\partial_x \varphi)\, dt dx \geq 0.$$

The apparent contradiction of the above definitions with the Second Law of Thermodynamics is soon solved by noticing that the physical entropies are concave functions while the ones we are using here are convex.

As a direct consequence of the Dominate Converge Theorem we can state the following result.

Theorem 6.2.4 *Let $\{u_\varepsilon\}_{\varepsilon>0}$ and u be functions defined on $[0, \infty) \times \mathbb{R}$ with values in $\mathbb{R}$. If*

(i) there exists $M > 0$ such that $\|u_\varepsilon\|_{L^\infty((0,\infty)\times\mathbb{R})} \leq M$ for every $\varepsilon > 0$;
(ii) $u \in L^\infty((0, \infty) \times \mathbb{R})$;
(iii) $u_\varepsilon \to u$ in $L^1_{loc}((0, \infty) \times \mathbb{R})$ as $\varepsilon \to 0$;
(iv) every u_ε is a entropy solution of (6.11);

then

$$u \text{ is a entropy solution of (6.11).}$$

A fundamental class of entropies are the ones introduced by Kružkov [12]

$$\eta(\xi) = |\xi - c|, \qquad q(\xi) = \text{sign}\,(\xi - c)\,(f(\xi) - f(c)), \qquad \xi \in \mathbb{R}, \qquad (6.21)$$

for every constant $c \in \mathbb{R}$.

Since the Kružkov entropies are not C^2 the following theorem is needed.

Theorem 6.2.5 *Let $u : [0, \infty) \times \mathbb{R} \to \mathbb{R}$ be a function. If*

$$u \in L^\infty_{loc}((0, \infty) \times \mathbb{R}),$$

then the following statements are equivalent

(i) u is an entropy solution of (6.11)–(6.12);
(ii) for every $c \in \mathbb{R}$ and every nonnegative test function $\varphi \in C^\infty(\mathbb{R}^2)$ with compact support

$$\int_0^\infty \int_\mathbb{R} (|u - c|\partial_t \varphi + \text{sign}\,(u - c)\,(f(u) - f(c))\partial_x \varphi)\, dt dx$$

$$(6.22)$$

$$+ \int_\mathbb{R} |u_0(x) - c|\varphi(0, x)dx \geq 0.$$

Remark 6.2.3 The set of the entropies

$$\{\eta \in C^2(\mathbb{R}); \eta \text{ convex}\}$$

is an infinite dimensional manifold. On the other hand the set of the Kružkov entropies

$$\{| \cdot -c|; c \in \mathbb{R}\}$$

is a one-dimensional manifold. Therefore the previous theorem says that if we have to verify that a function is an entropy solution of (6.11) we can use just the Kružkov entropies and the "amount" of inequalities to verify is "much lower" than the one required in Definition 6.2.3.

Proof (of Theorem 6.2.5) Let us start by proving $(i) \Rightarrow (ii)$. Let $c \in \mathbb{R}$ and $\varphi \in C^\infty(\mathbb{R}^2)$ be a nonnegative test function with compact support. For every $n \in \mathbb{N} \setminus \{0\}$, consider the functions

$$\eta_n(\xi) = \sqrt{(\xi - c)^2 + \frac{1}{n}}, \qquad q_n(\xi) = \int_c^\xi \frac{\sigma - c}{\sqrt{(\sigma - c)^2 + \frac{1}{n}}} f'(\sigma) d\sigma, \qquad \xi \in \mathbb{R}.$$

Since

$$\eta_n \in C^2(\mathbb{R}),$$

$$\eta_n'(\xi) = \frac{\xi - c}{\sqrt{(\xi - c)^2 + \frac{1}{n}}},$$

$$\eta_n''(\xi) = \frac{1}{n\left((\xi - c)^2 + \frac{1}{n}\right)^{\frac{3}{2}}} \geq 0,$$

$$q_n' = \eta_n' f',$$

we have

$$\int_0^\infty \int_{\mathbb{R}} (\eta_n(u)\partial_t \varphi + q_n(u)\partial_x \varphi) \, dt dx + \int_{\mathbb{R}} \eta_n(u_0(x))\varphi(0, x) dx \geq 0.$$

As $n \to \infty$ thanks to the Dominated Convergence Theorem we get (6.22).

Let us prove $ii) \Rightarrow i)$. Let η be an entropy with flux q and $\varphi \in C^\infty(\mathbb{R}^2)$ be a nonnegative test function with compact support. Define

$$M = \sup_{\text{supp}(\varphi)} |u|.$$

We approximate η' with piecewise constant functions in $[-M, M]$. For every $n \in \mathbb{N} \setminus \{0\}$ consider

$$\eta_n(\xi) = \int_{-M}^{\xi} k_n(\sigma)d\sigma + \eta(-M),$$

$$k_n(\sigma) = \sum_{j=0}^{2n-1} \eta'\left(\frac{M}{n}j - M\right) \chi_{\left[\frac{M}{n}j - M, \frac{M}{n}(j+1) - M\right]}(\sigma),$$

$$q_n(\xi) = \int_{-M}^{\xi} f'(\sigma)k_n(\sigma)d\sigma.$$

We have

$$k_n(\sigma) = \sum_{j=0}^{n-1} a_j \left[\text{sign}\left(\sigma - b_j\right) + c_j\right] \chi_{\left[2\frac{M}{n}j - M, 2\frac{M}{n}(j+1) - M\right]}(\sigma),$$

where

$$a_j = \frac{1}{2}\left(\eta'\left(\frac{M}{n}(2j+1) - M\right) - \eta'\left(\frac{M}{n}2j - M\right)\right),$$

$$b_j = \frac{M}{n}(2j+1) - M,$$

$$c_j = \frac{1}{2}\left(\eta'\left(\frac{M}{n}(2j+1) - M\right) + \eta'\left(\frac{M}{n}2j - M\right)\right).$$

Since $\eta'' \geq 0$ we have $a_j \geq 0$ and then

$$\int_0^{\infty} \int_{\mathbb{R}} (\eta_n(u)\partial_t\varphi + q_n(u)\partial_x\varphi) \, dt dx + \int_{\mathbb{R}} \eta_n(u_0(x))\varphi(0, x)dx \geq 0.$$

As $n \to \infty$ thanks to the Dominated Convergence Theorem we get (6.19). $\qquad \square$

It is clear that a smooth solutions is both an entropy and a weak solution (see Remark 6.2.2). We conclude this section proving that the entropy solutions are weak solutions. In the next section we will show that there are weak solutions that are not entropy ones.

Theorem 6.2.6 *Let $u : [0, \infty) \times \mathbb{R} \to \mathbb{R}$ be a function. If*

$$u \in L_{loc}^{\infty}((0, \infty) \times \mathbb{R})$$

and u is an entropy solution of (6.11)–(6.12), then u is a weak solution of (6.11)–(6.12).

Proof Let $\varphi \in C^2(\mathbb{R}^2)$ be a test function with compact support. Define

$$\varphi_+ = \max\{\varphi, 0\}, \qquad \varphi_- = \max\{-\varphi, 0\},$$

clearly

$$\varphi = \varphi_+ - \varphi_-, \qquad \varphi_+, \varphi_- \geq 0.$$

Using a smooth approximation of $\varphi_\pm$ and then passing to the limit we get

$$\int_0^\infty \int_{\mathbb{R}} (|u - c|\partial_t \varphi_\pm + \text{sign}\,(u - c)\,(f(u) - f(c))\partial_x \varphi_\pm)\,dt dx$$

$$+ \int_{\mathbb{R}} |u_0(x) - c|\varphi_\pm(0, x)dx \geq 0, \tag{6.23}$$

for every $c \in \mathbb{R}$.

Define

$$M = \sup_{\text{supp}(\varphi)} |u|.$$

Choosing $c = M + 1$ in (6.23) we get

$$\int_0^\infty \int_{\mathbb{R}} ((M + 1 - u)\partial_t \varphi_\pm + (f(M + 1) - f(u))\partial_x \varphi_\pm)\,dt dx$$

$$+ \int_{\mathbb{R}} (M + 1 - u_0(x))\varphi_\pm(0, x)dx \geq 0,$$

and integrating by parts (since $M + 1$ is a classical solution of (6.11)) we get

$$\int_0^\infty \int_{\mathbb{R}} (u\partial_t \varphi_\pm + f(u)\partial_x \varphi_\pm)\,dt dx + \int_{\mathbb{R}} u_0(x)\varphi_\pm(0, x)dx \leq 0. \tag{6.24}$$

On the other hand, if we choose $c = -M - 1$ in (6.23) we get

$$\int_0^\infty \int_{\mathbb{R}} ((u + M + 1)\partial_t \varphi_\pm + (f(u) - f(-M - 1))\partial_x \varphi_\pm)\,dt dx$$

$$+ \int_{\mathbb{R}} (u_0(x) + M + 1)\varphi_\pm(0, x)dx \geq 0,$$

and integrating by parts (since $-M - 1$ is a classical solution of (6.11)) we get

$$\int_0^\infty \int_\mathbb{R} (u\partial_t \varphi_\pm + f(u)\partial_x \varphi_\pm)\, dt dx + \int_\mathbb{R} u_0(x)\varphi_\pm(0, x)dx \geq 0. \qquad (6.25)$$

Adding (6.24) and (6.25) we get (6.19).   □

6.2.5 Entropic Shocks

In Sect. 6.2.2 we introduced the shock U (see (6.14)) and proved that it is a weak solution of (6.11) if and only if the Rankine-Hugoniot Condition (6.15) holds. In this section we prove a similar result giving a necessary and sufficient condition for the shock to be an entropy solution.

Theorem 6.2.7 *The following statements are equivalent:*

(i) *the function U defined in (6.14) is an entropy solution of (6.11);*
(ii) *the Rankine-Hugoniot Condition holds true, i.e.,*

$$f(u_+) - f(u_-) = \lambda(u_+ - u_-), \qquad (6.26)$$

and

$$\begin{cases} f(\theta u_+ + (1 - \theta)u_-) \geq \theta f(u_+) + (1 - \theta)f(u_-), & \text{if } u_- < u_+, \\ f(\theta u_+ + (1 - \theta)u_-) \leq \theta f(u_+) + (1 - \theta)f(u_-), & \text{if } u_- > u_+, \end{cases} \qquad (6.27)$$

for every $0 < \theta < 1$.

The inequalities in (6.27) have a simple geometric interpretation. If $u_- < u_+$ the graph of f has to be above the segment connecting $(u_-, f(u_-))$ and $(u_+, f(u_+))$, that is always true if f is concave. On the other hand if $u_- > u_+$ the graph of f has to be below the segment connecting $(u_+, f(u_+))$ and $(u_-, f(u_-))$, that is always trues if f is convex. In particular, if f is concave the entropic shocks are upward and if is convex they are downward.

Moreover, we can rewrite (6.27) in the following way

$$\frac{f(u_*) - f(u_-)}{u_* - u_-} \geq \frac{f(u_+) - f(u_*)}{u_+ - u_*}, \qquad (6.28)$$

for every $\min\{u_+, u_-\} < u_* < \max\{u_+, u_-\}$.

Indeed, if $u_- < u_+$ (in the case $u_- > u_+$ the same argument works) and $u_* = \theta u_+ + (1 - \theta)u_-$ for some $0 < \theta < 1$ we have

$$\frac{f(u_*) - f(u_-)}{u_* - u_-} - \frac{f(u_+) - f(u_*)}{u_+ - u_*}$$

$$= \frac{f(u_*)(u_+ - u_-)}{(u_* - u_-)(u_+ - u_*)} - \frac{f(u_-)}{u_* - u_-} - \frac{f(u_+)}{u_+ - u_*}$$

$$\geq \frac{(\theta f(u_+) + (1 - \theta)f(u_-))(u_+ - u_-)}{(u_* - u_-)(u_+ - u_*)} - \frac{f(u_-)}{u_* - u_-} - \frac{f(u_+)}{u_+ - u_*}$$

$$= f(u_+)\frac{\theta(u_+ - u_-) - (u_* - u_-)}{(u_* - u_-)(u_+ - u_*)} + f(u_-)\frac{(1 - \theta)(u_+ - u_-) - (u_+ - u_*)}{(u_* - u_-)(u_+ - u_*)} = 0.$$

Let us observe that (6.28) represents a stability condition. Indeed, if $u_- < u_* < u_+$ we can perturb the shock (u_-, u_+) and split it in the two shocks (u_-, u_*), (u_*, u_+). The two quantities in (6.28) give the speed of these two shocks: the one on the left is faster than the one on the right. Then the two waves will interact in finite time and generate again the initial shock (u_-, u_+) (see Fig. 6.3).

Lemma 6.2.1 *The following statements are equivalent:*

(i) the function U defined in (6.14) is an entropy solution of (6.11);
(ii) for every entropy η with flux q the following inequlity holds

$$\lambda(\eta(u_+) - \eta(u_-)) \geq q(u_+) - q(u_-); \tag{6.29}$$

(iii) for every constant $c \in \mathbb{R}$

$$\lambda(|u_+ - c| - |u_- - c|)$$
$$\geq \text{sign}(u_+ - c)(f(u_+) - f(c)) \tag{6.30}$$
$$- \text{sign}(u_- - c)(f(u_-) - f(c)).$$

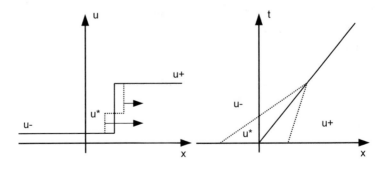

Fig. 6.3 Shock wave (u_-, u_+) with $u_- < u_+$

Proof Let $\varphi \in C^\infty((0, \infty) \times \mathbb{R})$ be a nonnegative test function with compact support and η be an entropy with flux q. Consider the vector field

$$G = (\eta(U)\varphi, q(U)\varphi)$$

and the domains

$$\Omega_+ = \{x > \lambda t\}, \qquad \Omega_- = \{x < \lambda t\}.$$

The definition of U gives

$$(t, x) \in \Omega_+ \implies \begin{cases} G(t, x) = (\eta(u_+)\varphi, q(u_+)\varphi), \\ \mathrm{div}_{(t,x)}(G)(t, x) = \eta(u_+)\partial_t\varphi + q(u_+)\partial_x\varphi, \end{cases}$$

$$(t, x) \in \Omega_- \implies \begin{cases} G(t, x) = (\eta(u_-)\varphi, q(u_-)\varphi), \\ \mathrm{div}_{(t,x)}(G)(t, x) = \eta(u_-)\partial_t\varphi + q(u_-)\partial_x\varphi. \end{cases}$$

Since

$$\partial\Omega_+ = \partial\Omega_- = \{x = \lambda t\},$$

and the outer normals to Ω_+ and Ω_- are $(\lambda, -1)$ and $(-\lambda, 1)$ we have

$$\int_0^\infty \int_\mathbb{R} (\eta(U)\partial_t\varphi + q(U)\partial_x\varphi)dtdx$$

$$= \iint_{\Omega_+} (\eta(u_+)\partial_t\varphi + q(u_+)\partial_x\varphi)dtdx + \iint_{\Omega_-} (\eta(u_-)\partial_t\varphi + q(u_-)\partial_x\varphi)dtdx$$

$$= \iint_{\Omega_+} \mathrm{div}(G)dtdx + \iint_{\Omega_-} \mathrm{div}(G)dtdx$$

$$= \int_0^\infty (\eta(u_+), q(u_+)) \cdot (\lambda, -1)\varphi(t, \lambda t)dt + \int_0^\infty (\eta(u_-), q(u_-)) \cdot (-\lambda, 1)\varphi(t, \lambda t)dt$$

$$= \left[\lambda(\eta(u_+) - \eta(u_-)) - (q(u_+) - q(u_-))\right] \int_0^\infty \varphi(t, \lambda t)dt.$$

Therefore

$$\int_0^\infty \int_\mathbb{R} (\eta(U)\partial_t\varphi + q(U)\partial_x\varphi)dtdx \geq 0, \quad \forall\varphi$$

$$\Updownarrow$$

$$\lambda(\eta(u_+) - \eta(u_-)) \geq q(u_+) - q(u_-).$$

Therefore we have proved that $i) \Leftrightarrow ii)$. The same argument works for $i) \Leftrightarrow iii)$.

$\square$

Proof (of Theorem 6.2.7) We begin by proving that $(i) \Rightarrow (ii)$. Since U is an entropy solution of (6.11), Theorem 6.2.2 gives (6.26). We have to prove (6.27). We distinguish two cases. We assume $u_- < u_+$. Let $0 < \theta < 1$ be fixed. We choose

$$c = \theta u_+ + (1 - \theta)u_-.$$

Since

$$u_- < c < u_+,$$

(6.30) gives

$$f(u_+) + f(u_-) - 2f(c) \leq \lambda(u_+ + u_- - 2c). \tag{6.31}$$

Using (6.26) and (6.31)

$$
\begin{aligned}
2f(\theta u_+ + (1 - \theta)u_-) &= 2f(c) \\
&\geq f(u_+) + f(u_-) - \lambda(u_+ + u_- - 2c) \\
&= f(u_+) + f(u_-) - \lambda(u_+ + u_- - 2(\theta u_+ + (1 - \theta)u_-)) \\
&= f(u_+) + f(u_-) - \lambda(1 - 2\theta)(u_+ - u_-) \\
&= f(u_+) + f(u_-) - (1 - 2\theta)(f(u_+) - f(u_-)) \\
&= 2(\theta f(u_+) + (1 - \theta)f(u_-)).
\end{aligned}
$$

Since the case $u_+ < u_+$ is analogous (6.27) is proved.

We have to prove that $(ii) \Rightarrow (i)$. If is enough to verify that (6.30) holds for every $c \in \mathbb{R}$. We distinguish four cases.

If

$$c \leq \min\{u_+, u_-\},$$

(6.26) gives

$$
\begin{aligned}
\lambda(|u_+ - c| - |u_- - c|) &= \lambda(u_+ - u_-) \\
&= f(u_+) - f(u_-) = (f(u_+) - f(c)) - (f(u_-) - f(c)) \\
&= \operatorname{sign}(u_+ - c)(f(u_+) - f(c)) - \operatorname{sign}(u_- - c)(f(u_-) - f(c)).
\end{aligned}
$$

If

$$c \geq \max\{u_+, u_-\},$$

the same argument applies.

If

$$u_- < c < u_+,$$

there exists $0 < \theta < 1$ such that

$$c = \theta u_+ + (1 - \theta)u_-.$$

(6.27) guarantees

$$f(c) \geq \theta f(u_+) + (1 - \theta)f(u_-),$$

then using (6.26)

$$
\begin{aligned}
\lambda(|u_+ - c| - |u_- - c|) &= \lambda(u_+ + -u_- - 2c) \\
&= \lambda(1 - 2\theta)(u_+ - u_-) = (1 - 2\theta)(f(u_+) - f(u_-)) \\
&= f(u_+) + f(u_-) - 2(\theta f(u_+) + (1 - \theta)f(u_-)) \\
&\geq f(u_+) + f(u_-) - 2f(c) \\
&= \text{sign}(u_+ - c)(f(u_+) - f(c)) - \text{sign}(u_- - c)(f(u_-) - f(c)).
\end{aligned}
$$

Finally, if

$$u_- < c < u_+,$$

the same argument works. Then (6.30) holds for every $c \in \mathbb{R}$. □

Theorem 6.2.8 *Let $u : [0, \infty) \times \mathbb{R} \to \mathbb{R}$, $\tau > 0$, $\xi \in \mathbb{R}$. If*

(i) $u \in L^\infty_{loc}((0, \infty) \times \mathbb{R})$;
(ii) u is an entropy solution of (6.11);
(iii) $\displaystyle\lim_{\varepsilon \to 0} \frac{1}{\varepsilon^2} \int_{-\varepsilon}^{\varepsilon} \int_{-\varepsilon}^{\varepsilon} |u(t + \tau, x + \xi) - U(t, x)| dt dx = 0$;

then (6.26) and (6.27) hold.

Proof For every $\mu > 0$ define

$$u_\mu(t, x) = u(\mu t + \tau, \mu x + \xi), \qquad t \geq -\frac{\tau}{\mu}, x \in \mathbb{R}.$$

Since u is a weak solution of (6.11), the same does u_μ. We claim that

$$u_\mu \longrightarrow U, \qquad f(u_\mu) \longrightarrow f(U), \qquad \text{in } L^1_{loc}((0, \infty) \times \mathbb{R}), \text{ as } \mu \to 0. \qquad (6.32)$$

Let $R > 0$ and $\mu < \tau/R$. Since

$$U(\mu t, \mu x) = U(t, x), \qquad t > 0,\ x \in \mathbb{R},$$

we get

$$\int_{-R}^{R} \int_{-R}^{R} |u_\mu(t, x) - U(t, x)| \, dt\, dx$$

$$= \frac{1}{\mu^2} \int_{-R\mu}^{R\mu} \int_{-R\mu}^{R\mu} |u(t + \tau, x + \xi) - U(t, x)| \, dt\, dx \longrightarrow 0,$$

namely

$$u_\mu \longrightarrow U, \qquad \text{in } L^1((-R, R) \times (-R, R)),\ \text{as } \mu \to 0.$$

Therefore the Dominated Convergence Theorem gives (6.32). Finally, Theorem 6.2.4 and (6.32) implies that U is a entropy solution of (6.11). Then, the claim follows from Theorem 6.2.7. □

Example 6.2.1 The function

$$u(t, x) = \begin{cases} -\frac{2}{3}\left(t + \sqrt{3x + t^2}\right) & \text{if } 4x + t^2 > 0, \\ 0 & \text{if } 4x + t^2 < 0 \end{cases} \tag{6.33}$$

is an entropy solution of the Cauchy problem

$$\begin{cases} u_t + \left(\frac{u^2}{2}\right)_x = 0, & t > 0,\ x \in \mathbb{R}, \\ u(0, x) = \begin{cases} -\frac{2}{\sqrt{3}}\sqrt{x} & \text{if } x > 0, \\ 0 & \text{if } x < 0. \end{cases} \end{cases} \tag{6.34}$$

Introduce the notation

$$u_-(t, x) = 0, \quad u_+(t, x) = -\frac{2}{3}\left(t + \sqrt{3x + t^2}\right), \quad \lambda(t) = -\frac{t^2}{4}, \quad f(\xi) = \frac{\xi^2}{2}.$$

Since

$$\partial_x u_+(t, x) = -\frac{1}{\sqrt{3x + t^2}},$$

$$\partial_t u_+(t, x) = -\frac{2}{3}\left(1 + \frac{t}{\sqrt{3x + t^2}}\right),$$

$$u_+(t, x)\partial_x u_+(t, x) = \frac{2}{3}\left(\frac{t}{\sqrt{3x + t^2}} + 1\right)$$

u_- and u_+ are a classical solution of the Burgers equation.

We have only to verify that (6.26) and (6.27) hold along the curve $x = \lambda(t)$. Since

$$u_-(t, \lambda(t)) = 0,$$

$$u_+(t, \lambda(t)) = -t \leq 0,$$

$$f(u_+(t, \lambda(t))) - f(u_-(t, \lambda(t))) - \lambda'(t)(u_+(t, \lambda(t)) - u_-(t, \lambda(t))) = 0,$$

the Rankine-Hugoniot Condition is satisfied and the jump is downward (note that f is convex).

6.2.6 Change of Coordinates

One of the features of the weak and entropy solutions is that they are not invariant under changes of coordinates. These ones transform smooth solutions in smooth solutions but in general they do not transform weak/entropy solutions in weak/entropy solutions. Let us consider the following simple example based on the Burgers equation. We know that the shock

$$u(t, x) = \begin{cases} 1, & \text{if } x < t/2, \\ 0 & \text{if } x \geq t/2 \end{cases} \tag{6.35}$$

provides an entropy solution of the Riemann problem

$$\partial_t u + \partial_x\left(\frac{u^2}{2}\right) = 0, \qquad u(0, x) = \begin{cases} 1, & \text{if } x < 0, \\ 0 & \text{if } x \geq 0. \end{cases} \tag{6.36}$$

Consider the new unknown

$$v = u^3.$$

(6.35) and (6.36) become

$$v(t, x) = \begin{cases} 1, & \text{if } x < t/2, \\ 0 & \text{if } x \geq t/2 \end{cases} \tag{6.37}$$

and

$$\partial_t v + \partial_x \left(\frac{3}{4} v^{4/3} \right) = 0, \qquad v(0, x) = \begin{cases} 1, & \text{if } x < 0, \\ 0 & \text{if } x \geq 0. \end{cases} \qquad (6.38)$$

respectively. Since v does not satisfy the Rankine-Hugoniot condition, it does not provide a weak solution of (6.38).

6.2.7 Uniqueness and Stability of Entropy Solutions

In this section we prove the classical Kružkov theorem [12]. It has three main consequences: the uniqueness of the entropy solutions, the L^1 Lipschitz continuity with respect to the initial condition of the entropy solutions, and the finite speed of propagation of the waves generated by conservation laws.

Theorem 6.2.9 (Kružkov [12]) *Let $u, v : [0, \infty) \times \mathbb{R} \to \mathbb{R}$ be two entropy solutions of (6.11). If*

$$u, v \in L^\infty((0, \infty) \times \mathbb{R}),$$

then

$$\int_{-R}^{R} |u(t_2, x) - v(t_2, x)| dx \leq \int_{-R-L(t_2-t_1)}^{R+L(t_2-t_1)} |u(t_1, x) - v(t_1, x)| dx, \qquad (6.39)$$

for every $R > 0$ and almost every $0 \leq t_1 \leq t_2$, where

$$L = \sup_{(0,\infty) \times \mathbb{R}} (|f'(u)| + |f'(v)|).$$

A fundamental consequence of Kružkov theorem is the following.

Corollary 1 (Uniqueness and Stability of Entropy Solutions) *Let $u, v : [0, \infty) \times \mathbb{R} \to \mathbb{R}$ be two entropy solutions of (6.11). If*

$$u, v \in L^\infty((0, \infty) \times \mathbb{R}),$$

$$u(0, \cdot) - v(0, \cdot) \in L^1(\mathbb{R}) \ (or \ u(0, \cdot), v(0, \cdot) \in L^1(\mathbb{R})),$$

then

$$u(t, \cdot) - v(t, \cdot) \in L^1(\mathbb{R}) \ (or \ u(t, \cdot), v(t, \cdot) \in L^1(\mathbb{R})),$$

$$\|u(t, \cdot) - v(t, \cdot)\|_{L^1(\mathbb{R})} \leq \|u(0, \cdot) - v(0, \cdot)\|_{L^1(\mathbb{R})}, \qquad (6.40)$$

for almost every t ≥ 0. In particular

$$u(0, \cdot) = v(0, \cdot) \Longrightarrow u = v.$$

The proof of the Kružkov theorem is based on the following lemma.

Lemma 6.2.2 (Doubling of Variables)   *Let* $u, v : [0, \infty) \times \mathbb{R} \to \mathbb{R}$ *be two entropy solutions of (6.11). If*

$$u, v \in L^{\infty}((0, \infty) \times \mathbb{R}),$$

then

$$\partial_t |u - v| + \partial_x \left(\text{sign} \, (u - v) \, (f(u) - f(v)) \right) \leq 0 \tag{6.41}$$

holds in the sense of distributions on $(0, \infty) \times \mathbb{R}$.

Proof   Let $\varphi = \varphi(t, s, x, y)$ be a C^{∞} nonnegative test function defined on $(0, \infty) \times (0, \infty) \times \mathbb{R} \times \mathbb{R}$. Since u and v are entropy solutions of (6.11) we have

$$\int_0^{\infty} \int_{\mathbb{R}} \left(|u(t, x) - v(s, y)| \partial_t \varphi(t, s, x, y) \right.$$

$$\left. + \text{sign} \, (u(t, x) - v(s, y)) \, (f(u(t, x)) - f(v(s, y))) \partial_x \varphi(t, s, x, y) \right) dt dx \geq 0,$$

$$\int_0^{\infty} \int_{\mathbb{R}} \left(|v(s, y) - u(t, x)| \partial_s \varphi(t, s, x, y) \right.$$

$$\left. + \text{sign} \, (v(s, y) - u(t, x)) \, (f(v(s, y)) - f(u(t, x))) \partial_y \varphi(t, s, x, y) \right) ds dy \geq 0,$$

and then

$$\int_0^{\infty} \int_0^{\infty} \int_{\mathbb{R}} \int_{\mathbb{R}} \left(|u(t, x) - v(s, y)| (\partial_t \varphi + \partial_s \varphi) \right.$$

$$+ \text{sign} \, (u(t, x) - v(s, y)) \times \tag{6.42}$$

$$\left. \times (f(u(t, x)) - f(v(s, y)))(\partial_x \varphi + \partial_y \varphi) \right) dt ds dx dy \geq 0.$$

Let $\psi \in C^{\infty}((0, \infty) \times \mathbb{R})$ be a nonnegative test function and $\delta \in C^{\infty}(\mathbb{R})$ be such that

$$\delta \geq 0, \qquad \|\delta\|_{L^1(\mathbb{R})} = 1, \qquad \text{supp}(\delta) \subset [-1, 1].$$

Define

$$\delta_n(x) = n\delta(nx),$$

$$\varphi_n(t, s, x, y) = \psi\left(\frac{t+s}{2}, \frac{x+y}{2}\right)\delta_n\left(\frac{s-t}{2}\right)\delta_n\left(\frac{y-x}{2}\right). \tag{6.43}$$

We use φ_n as test function in (6.42)

$$\int_0^\infty\int_0^\infty\int_{\mathbb{R}}\int_{\mathbb{R}}\delta_n\left(\frac{s-t}{2}\right)\delta_n\left(\frac{y-x}{2}\right)\left((|u(t,x)-v(s,y)|\partial_t\psi\left(\frac{t+s}{2}, \frac{x+y}{2}\right)\right.$$

$$+ \,\text{sign}\,(u(t,x)-v(s,y)) \times$$

$$\left.\times\, (f(u(t,x))-f(v(s,y)))\partial_x\psi\left(\frac{t+s}{2}, \frac{x+y}{2}\right)\right)dtdsdxdy \geq 0.$$

As $n \to \infty$ we get

$$\int_0^\infty\int_{\mathbb{R}}(|u-v|\partial_t\psi + \text{sign}\,(u-v)\,(f(u)-f(v))\partial_x\psi)\,dtdx \geq 0,$$

that gives the claim. $\square$

Proof (of Theorem 6.2.9) Let $R > 0$ and $0 \leq t_1 \leq t_2$. Define

$$\alpha_n(x) = \int_{-\infty}^x \delta_n(y)dy, \qquad x \in \mathbb{R},$$

where δ_n is defined in (6.43). Consider the test function

$$\varphi_n(t, x) = (\alpha_n(t-t_1) - \alpha_n(t-t_2))\left(1 - \alpha_n\left(\sqrt{x^2 + \frac{1}{n}} - R - L(t_2 - t)\right)\right),$$

that is a smooth approximant of the characteristic function of the set

$$\left\{(t, x) \in [0, \infty) \times \mathbb{R}; t_1 \leq t \leq t_2, |x| \leq R + L(t_2 - t)\right\}.$$

Testing (6.41) with φ_n we get

$$\int_0^\infty\int_{\mathbb{R}}|u-v|\,(\delta_n(t-t_1) - \delta_n(t-t_2))\left(1 - \alpha_n\left(\sqrt{x^2 + \frac{1}{n}} - R - L(t_2 - t)\right)\right)dtdx$$

$$- L\int_0^\infty\int_{\mathbb{R}}|u-v|\,(\alpha_n(t-t_1) - \alpha_n(t-t_2))\delta_n\left(\sqrt{x^2 + \frac{1}{n}} - R - L(t_2 - t)\right)dtdx$$

$$+ \int_0^\infty \int_{\mathbb{R}} \text{sign} (u - v) (f(u) - f(v)) (\alpha_n(t - t_1) - \alpha_n(t - t_2)) \cdot$$

$$\cdot \frac{x}{\sqrt{x^2 + \frac{1}{n}}} \delta_n \left(\sqrt{x^2 + \frac{1}{n}} - R - L(t_2 - t) \right) dt dx \geq 0.$$

Since

$$|f(u) - f(v)| \leq |u - v|, \qquad \left| \frac{x}{\sqrt{x^2 + \frac{1}{n}}} \right| \leq 1$$

we have

$$\int_0^\infty \int_{\mathbb{R}} |u - v| (\delta_n(t - t_1) - \delta_n(t - t_2)) \left(1 - \alpha_n \left(\sqrt{x^2 + \frac{1}{n}} - R - L(t_2 - t) \right) \right) dt dx$$

$$\geq \int_0^\infty \int_{\mathbb{R}} \left(L|u - v| - \text{sign} (u - v) (f(u) - f(v)) \frac{x}{\sqrt{x^2 + \frac{1}{n}}} \right) \cdot$$

$$\cdot (\alpha_n(t - t_1) - \alpha_n(t - t_2)) \delta_n \left(\sqrt{x^2 + \frac{1}{n}} - R - L(t_2 - t) \right) dt dx \geq 0.$$

As $n \to \infty$, using the fact that, due to the Lusin Theorem, the map $t \geq 0 \mapsto u(t, \cdot) - v(t, \cdot) \in L^1_{loc}(\mathbb{R})$ is almost everywhere continuous, we get (6.39). $\qquad \square$

6.3 Riemann Problem

In Sect. 6.2.7 we proved the uniqueness and stability of entropy solutions of Cauchy problems. Here we focus on the existence of entropy solutions. We analyze the simplest cases: the Riemann problems, that are Cauchy problems with Heaviside type initial condition

$$\begin{cases} \partial_t u + \partial_x f(u) = 0, & t > 0, \; x \in \mathbb{R}, \\ u(0, x) = \begin{cases} u_+, & \text{if } x \geq 0, \\ u_-, & \text{if } x < 0, \end{cases} \end{cases} \qquad (6.44)$$

where $f \in C^2(\mathbb{R})$ and $u_- \neq u_+$ are constants.

In the following sections we first consider the case in which f is convex. Indeed the solutions obtained under that assumption are the building blocks of the solutions of the general case [5, 6, 11].

6.3.1 Strictly Convex Fluxes

We assume that f is a convex function, the concave case is analogous.
    We distinguish two cases. If (see Fig. 6.4)

$$u_+ < u_-$$

then the entropy solution of (6.44) is the *shock wave* (see Fig. 6.5)

$$u(t, x) = \begin{cases} u_+, & \text{if } x \geq \dfrac{f(u_+) - f(u_-)}{u_+ - u_-}t, \\[3mm] u_-, & \text{if } x < \dfrac{f(u_+) - f(u_-)}{u_+ - u_-}t. \end{cases}$$

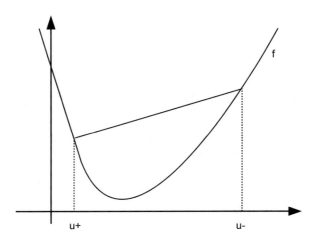

Fig. 6.4 Convex flux f

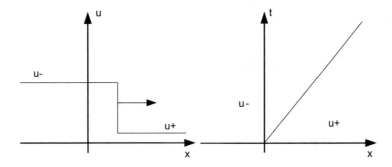

Fig. 6.5 Shck wave (u_-, u_+) with $u_+ < u_-$

If (see Fig. 6.6)

$$u_+ > u_-$$

then the entropy solution of (6.44) is the *rarefaction wave* (see Fig. 6.7)

$$u(t, x) = \begin{cases} u_+, & \text{if } x \geq f'(u_+)t, \\ \sigma, & \text{if } x = f'(\sigma)t, \ u_- < \sigma < u_+, \\ u_-, & \text{if } x = f'(u_-)t. \end{cases} \tag{6.45}$$

Observe that the definition makes sense because f is convex and then f' is increasing.

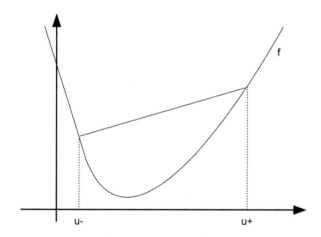

Fig. 6.6 Shock wave (u_-, u_+) with $u_+ > u_-$

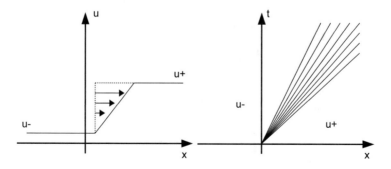

Fig. 6.7 Rarefaction wave

We claim that

$$\partial_t \eta(u) + \partial_x q(u) = 0, \tag{6.46}$$

for every entropy η with flux q, where u is the rarefaction wave defined in (6.45).
Consider the sets

$$\Omega_1 = \{(t, x) \in (0, \infty) \times \mathbb{R}; x < f'(u_-)t\},$$
$$\Omega_2 = \{(t, x) \in (0, \infty) \times \mathbb{R}; f'(u_-)t < x < f'(u_+)t\},$$
$$\Omega_3 = \{(t, x) \in (0, \infty) \times \mathbb{R}; x > f'(u_+)t\},$$

whit outer normals n_1, n_2, n_3, and a nonnegative test function $\varphi \in C^\infty((0, \infty) \times \mathbb{R})$. We have

$$\int_0^\infty \int_\mathbb{R} (\eta(u)\partial_t \varphi + q(u)\partial_x \varphi) dt dx$$

$$= -\sum_{i=1}^3 \iint_{\Omega_i} \underbrace{(\partial_t \eta(u) + \partial_x q(u))}_{=0 \text{ (because } u \text{ is smooth in each } \Omega_i)} \varphi dt dx$$

$$+ \sum_{i=1}^3 \int_{\partial\Omega_i} \underbrace{(\eta(u)\varphi, q(u)\varphi) \cdot n_i d\sigma = 0.}_{=0 \text{ (because } u \text{ is continuous)}}$$

Therefore (6.46) holds and then (6.45) is the entropy solution of (6.44).

When f is concave we have a completely symmetric case, a shock when $u_- < u_+$ and a rarefaction when $u_- > u_+$.

Example 6.3.1 The entropy solution of the Riemann problem

$$\begin{cases} u_t + \left(\frac{u^2}{2}\right)_x = 0, & t > 0, \ x \in \mathbb{R}, \\ u(0, x) = \begin{cases} -1 & \text{if } x < 0, \\ 1 & \text{if } x \geq 0, \end{cases} \end{cases}$$

is the rarefaction wave

$$u(t, x) = \begin{cases} -1 & \text{if } x < -t, \\ \sigma & \text{if } x = \sigma t, \ -1 < \sigma \leq 1, \\ 1 & \text{if } x > t. \end{cases}$$

Example 6.3.2 The entropy solution of the Riemann problem

$$\begin{cases} u_t + (u^3)_x = 0, & t > 0, \ x \in \mathbb{R}, \\ u(0, x) = \begin{cases} 1 & \text{if } x < 2, \\ 0 & \text{if } x \geq 2, \end{cases} \end{cases}$$

is the shock

$$u(t, x) = \begin{cases} 1 & \text{if } x < t + 2, \\ 0 & \text{if } x \geq t + 2. \end{cases}$$

Example 6.3.3 The entropy solution of the Riemann problem

$$\begin{cases} u_t + (u^3)_x = 0, & t > 0, \ x \in \mathbb{R}, \\ u(0, x) = \begin{cases} 0 & \text{if } x < 2, \\ 1 & \text{se } x \geq 2, \end{cases} \end{cases}$$

is the rarefaction wave

$$u(t, x) = \begin{cases} 0 & \text{if } x < 2, \\ \sigma & \text{if } x = 3\sigma^2 t + 2, \ 0 < \sigma \leq 1, \\ 1 & \text{if } x > 3t + 2. \end{cases}$$

Example 6.3.4 The entropy solution of the Riemann problem

$$\begin{cases} u_t + (e^u)_x = 0, & t > 0, \ x \in \mathbb{R}, \\ u(0, x) = \begin{cases} 2 & \text{if } x < 0, \\ 0 & \text{if } x \geq 0, \end{cases} \end{cases}$$

is the shock

$$u(t, x) = \begin{cases} 2 & \text{if } x < \frac{e^2 - 1}{2} t, \\ 0 & \text{if } x \geq \frac{e^2 - 1}{2} t. \end{cases}$$

6.3.2 General Fluxes

In the case of convex or concave fluxes the solution of the Riemann problem (6.44) consists of only one wave, a shock or a rarefaction wave. In the case of fluxes that are not convex or concave we can have several waves of both types. Moreover, the waves may also be glued together.

We have to distinguish again two cases. If

$$u_- < u_+$$

we consider the convex hull f_* of f in the interval $[u_-, u_+]$, i.e., f_* is the largest convex map such that

$$f_*(\xi) \le f(\xi), \qquad u_- \le \xi \le u_+.$$

Let consider the points $w_0, \ldots, w_n$ such that (see Fig. 6.8)

$$u_- = w_0 < w_1 < \ldots < w_n = u_+,$$

$$f(w_i) = f_*(w_i), \quad i = 0, \ldots, n,$$

$$w_i < u < w_{i+1} \Rightarrow f_*(u) < f(u) \text{ or } f_*(u) = f(u), \quad i = 0, \ldots, n-1.$$

We solve separately the $n-1$ Riemann problems obtained in correspondence of the values (w_i, w_{i+1}), $i = 0, \ldots, n-1$. If $f < f_*$ in (w_i, w_{i+1}) we have a shock otherwise a rarefaction (see Fig. 6.9). This algorithm provides clearly the entropy solution of (6.44) because we are gluing entropy solutions.

If

$$u_- > u_+$$

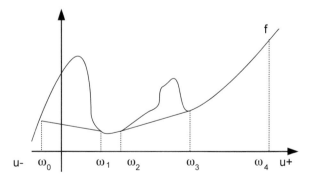

Fig. 6.8 Nonconvex flux f

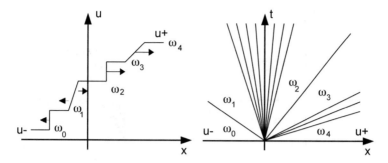

Fig. 6.9 Nonconvex flux with shock (u_-, u_+) with $u_- < u_+$

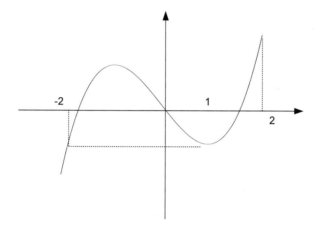

Fig. 6.10 $f(u) = (u^3 - 3u)$

we consider the concave hull f^* of f in the interval $[u_-, u_+]$, i.e., f^* is the smallest concave map such that

$$f(\xi) \leq f^*(\xi), \qquad u_- \leq \xi \leq u_+,$$

and we argue in the same way.

Example 6.3.5 Consider the Riemann problem (see Fig. 6.10)

$$\begin{cases} \partial_t u + \partial_x (u^3 - 3u) = 0, & t > 0, \ x \in \mathbb{R}, \\ u(0, x) = \begin{cases} 2, & \text{if } x \geq 0, \\ -2, & \text{if } x < 0, \end{cases} \end{cases} \tag{6.47}$$

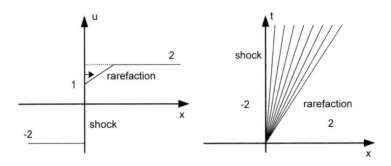

Fig. 6.11 Solution of (6.47)

The solution of (6.47) is (see Fig. 6.11)

$$u(0, x) = \begin{cases} 2, & \text{if } x \geq 9t, \\ \sigma, & \text{if } x = (3\sigma^2 - 3)t,\ 1 \leq \sigma < 2, \\ -2, & \text{if } x < 0, \end{cases}$$

where the shock connecting -2 and 1 is attached to the rarefaction from 1 to 2.

The same feature can be found in

$$\begin{cases} \partial_t u + \partial_x (u^3 - 3u) = 0, & t > 0,\ x \in \mathbb{R}, \\ u(0, x) = \begin{cases} -2, & \text{if } x \geq 0, \\ 2, & \text{if } x < 0. \end{cases} \end{cases}$$

Example 6.3.6 Let us solve the Cauchy problem

$$\begin{cases} u_t + \left(\frac{u^2}{2}\right)_x = 0, & t > 0,\ x \in \mathbb{R}, \\ u(0, x) = \begin{cases} 1 & \text{if } 0 < x < 1 \\ 0 & \text{otherwise.} \end{cases} \end{cases} \tag{6.48}$$

The wave generated at $x = 0$ is a rarefaction wave with speeds between 0 and 1, the one generated at $x = 1$ a shock with speed 1/2, they interact at $t = 2$, and we have (see Fig. 6.12)

$$u(t, x) = \begin{cases} 0, & \text{if } x \leq 0, \\ \sigma, & \text{if } x = \sigma t,\ 0 \leq \sigma \leq 1, \\ 1, & \text{if } t < x \leq \frac{t}{2} + 1, \\ 0, & \text{if } x > \frac{t}{2} + 1, \end{cases} \qquad 0 \leq t \leq 2. \tag{6.49}$$

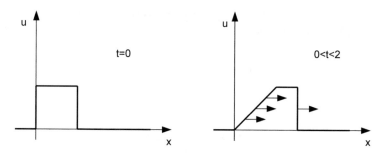

Fig. 6.12 Solution of (6.48)

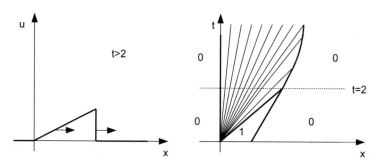

Fig. 6.13 Solution of (6.48)

For $t \geq 2$ we have a structure of the type (see Fig. 6.13)

$$u(t, x) = \begin{cases} 0, & \text{if } x \leq 0, \\ \sigma, & \text{if } x = \sigma t,\ 0 \leq \sigma \leq \lambda(t), \qquad t \geq 2. \\ 0, & \text{if } x > \lambda(t), \end{cases} \qquad (6.50)$$

We have to determine $\lambda(t)$. We know that

$$\lambda(2) = 2. \qquad (6.51)$$

The Rankine-Hugoniot condition gives

$$\lambda'(t) = \frac{u(t, \lambda(t)^-)}{2}. \qquad (6.52)$$

Finally, from (6.50) we know

$$u(t, \lambda(t)^-) = \frac{\lambda(t)}{t}. \qquad (6.53)$$

Therefore, (6.51), (6.52), and (6.53) imply that $\lambda(t)$ is the unique solution of the ordinary differential problem

$$\lambda'(t) = \frac{\lambda(t)}{2t}, \qquad \lambda(2) = 2,$$

namely

$$\lambda(t) = \sqrt{2t}, \qquad t \geq 2.$$

Example 6.3.7 Let us solve the Cauchy problem

$$\begin{cases} u_t + \left(\frac{u^2}{2}\right)_x = 0, & t > 0, \ x \in \mathbb{R}, \\[2mm] u(0, x) = \begin{cases} 1 & \text{se } x < -1, \\ 0 & \text{se } -1 < x < 0 \\ 2 & \text{se } 0 < x < 1 \\ 0 & \text{se } x > 1. \end{cases} \end{cases} \tag{6.54}$$

The wave generated at $x = -1$ is a shock with speed $1/2$, the one generated at $x = 0$ is a rarefaction wave with speeds between 0 and 2, the one generated at $x = 1$ a shock with speed 1. The first interaction is between the second and the third wave at $t = 1$, and we have (see Fig. 6.14).

$$u(t, x) = \begin{cases} 1, & \text{if } x \leq \frac{t}{2} - 1, \\ 0, & \text{if } \frac{t}{2} - 1 \leq x \leq 0, \\ \sigma, & \text{if } x = \sigma t, \ 0 \leq \sigma \leq 2, \qquad 0 \leq t \leq 1. \\ 2, & \text{if } 2t < x \leq t + 1, \\ 0, & \text{if } x > t + 1, \end{cases} \tag{6.55}$$

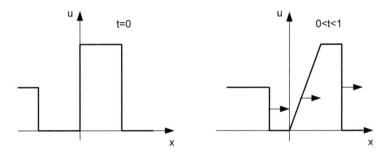

Fig. 6.14 Solution of (6.54)

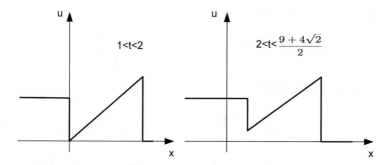

Fig. 6.15 Solution of (6.54)

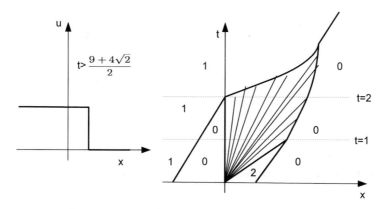

Fig. 6.16 Solution of (6.54)

The second interaction is between the first and the second wave at $t = 2$, and for $1 \leq t \leq 2$ and $t \geq 2$ we have a structure of the type (see Figs. 6.15 and 6.16)

$$u(t, x) = \begin{cases} 1, & \text{if } x \leq 0, \\ \sigma, & \text{if } x = \sigma t, \ 0 \leq \sigma \leq \lambda(t), \qquad 1 \leq t \leq 2, \\ 0, & \text{if } x > \lambda(t), \end{cases} \qquad (6.56)$$

$$u(t, x) = \begin{cases} 1, & \text{if } x \leq \gamma(t), \\ \sigma, & \text{if } x = \sigma t, \ \gamma(t) \leq \sigma \leq \lambda(t), \qquad t \geq 2. \\ 0, & \text{if } x > \lambda(t), \end{cases} \qquad (6.57)$$

We have to determine $\gamma(t)$ and $\lambda(t)$. We know that

$$\gamma(2) = 0, \qquad \lambda(1) = 2. \qquad (6.58)$$

The Rankine-Hugoniot condition gives

$$\gamma'(t) = \frac{1 + u(t, \gamma(t)^+)}{2}, \qquad \lambda'(t) = \frac{u(t, \lambda(t)^-)}{2}. \tag{6.59}$$

Finally, from (6.56) we know

$$u(t, \gamma(t)^+) = \frac{\gamma(t)}{t}, \qquad u(t, \lambda(t)^-) = \frac{\lambda(t)}{t}. \tag{6.60}$$

Therefore, (6.51), (6.52), and (6.53) imply that $\gamma(t)$ and $\lambda(t)$ are the unique solution of the ordinary differential problems

$$\begin{cases} \gamma'(t) = \dfrac{1}{2}\left(1 + \dfrac{\gamma(t)}{t}\right), \\ \gamma(2) = 0, \end{cases} \qquad \begin{cases} \lambda'(t) = \dfrac{\lambda(t)}{2t} \\ \lambda(1) = 2, \end{cases}$$

namely

$$\gamma(t) = t - \sqrt{2t}, \qquad \lambda(t) = 2\sqrt{t}.$$

Since, γ and λ interact at $\frac{9+4\sqrt{2}}{2}$, (6.57) holds only for $2 \le t \le \frac{9+4\sqrt{2}}{2}$. For $t \ge \frac{9+4\sqrt{2}}{2}$ we have only a shock connecting 0 and 1 with speed $\frac{1}{2}$

$$u(t, x) = \begin{cases} 1, & \text{if } x \le \frac{t}{2} + \sqrt{18 + 8\sqrt{2}}, \\ 0, & \text{if } x > \frac{t}{2} + \sqrt{18 + 8\sqrt{2}}, \end{cases} \qquad t \ge \frac{9 + 4\sqrt{2}}{2}.$$

6.4   Vanishing Viscosity

In this section we discuss the parabolic approximation

$$\begin{cases} \partial_t u_\varepsilon + \partial_x f(u_\varepsilon) = \varepsilon \partial_{xx}^2 u_\varepsilon, & t > 0, \ x \in \mathbb{R}, \\ u_\varepsilon(0, x) = u_{0,\varepsilon}(x), & x \in \mathbb{R}, \end{cases} \tag{6.61}$$

of the scalar hyperbolic conservation law

$$\begin{cases} \partial_t u + \partial_x f(u) = 0, & t > 0, \ x \in \mathbb{R}, \\ u(0, x) = u_0(x), & x \in \mathbb{R}. \end{cases} \tag{6.62}$$

The mean feature of such an approximation relies in the regularity property of the solutions. Indeed due to its parabolic structure (6.61) does not experience shocks.

For the initial data of (6.62) we assume

$$u_0 \in L^1(\mathbb{R}) \cap BV(\mathbb{R}).$$

On the other hand, for every $\varepsilon > 0$, $u_{0,\varepsilon}$ is a smooth approximation to u_0 such that

$$u_{0,\varepsilon} \in C^\infty(\mathbb{R}) \cap W^{2,1}(\mathbb{R}), \qquad \varepsilon > 0,$$

$$u_{0,\varepsilon} \longrightarrow u_0, \qquad \text{in } L^p(\mathbb{R}), \ 1 \le p < \infty, \text{ as } \varepsilon \to 0,$$

$$\|u_{0,\varepsilon}\|_{L^\infty(\mathbb{R})} \le \|u_0\|_{L^\infty(\mathbb{R})}, \qquad \|u_{0,\varepsilon}\|_{L^1(\mathbb{R})} \le \|u_0\|_{L^1(\mathbb{R})}, \qquad \varepsilon > 0, \qquad (6.63)$$

$$\|\partial_x u_{0,\varepsilon}\|_{L^1(\mathbb{R})} \le TV(u_0), \qquad \varepsilon \left\|\partial^2_{xx} u_{0,\varepsilon}\right\|_{L^1(\mathbb{R})} \le C, \qquad \varepsilon > 0,$$

for some constant $C > 0$ independent on ε. Under these assumptions (6.61) admits a unique solution u_ε such that [7, 14]

$$u_\varepsilon \in C^\infty([0, \infty) \times \mathbb{R}) \cap W^{2,p}((0, \infty); W^{1,p}(\mathbb{R})), \qquad 1 \le p < \infty.$$

The main result of this Section is the following [6, 11, 18].

Theorem 6.4.1 *If*

$$u_0 \in L^1(\mathbb{R}) \cap BV(\mathbb{R}).$$

then

$$u_\varepsilon \longrightarrow u \qquad \text{in } L^p_{loc}((0, \infty) \times \mathbb{R}), \ 1 \le p < \infty, \text{ and a.e. in } (0, \infty) \times \mathbb{R}, \qquad (6.64)$$

where u is the entropy weak solution of (6.62) and u_ε is the solution of (6.61). Moreover, the following estimate holds

$$\|u_\varepsilon(t, \cdot) - u(t, \cdot)\|_{L^1(\mathbb{R})} \le c\sqrt{\varepsilon t} \, TV(u_0) + \|u_{0,\varepsilon} - u_0\|_{L^1(\mathbb{R})}, \qquad (6.65)$$

for every $\varepsilon > 0$ and $t \ge 0$, where c is a positive constant independent on ε and t.

The convergence part of this result has been proved in [12] for scalar equations and in [4] for systems of conservation laws. The error estimates has been proved in [13].

Let us conclude this introduction with the following observation. In our statement all the family $\{u_\varepsilon\}_{\varepsilon > 0}$ converges to u and not just a subsequences, this result is due to the uniqueness of the entropy solutions of (6.62) and to the following equivalence

$$u_\varepsilon \longrightarrow u$$

$$\Updownarrow \qquad\qquad\qquad (6.66)$$

$$\forall \{u_{\varepsilon_k}\}_{k \in \mathbb{N}} \text{ subsequence } \exists \{u_{\varepsilon_{k_h}}\}_{h \in \mathbb{N}} \text{ subsequence s.t. } u_{\varepsilon_{k_h}} \longrightarrow u.$$

6.4.1 A Priori Estimates, Compactness, and Convergence

The aim of this section relies essentially in the proof of (6.64). Let us start with a technical lemma that will play a key role in the following a priori estimates.

Lemma 6.4.1 ([2, Lemma 2]) *Let* $v : \mathbb{R} \to \mathbb{R}$ *be a function. If*

$$v \in C^1(\mathbb{R}), \qquad v' \in L^1(\mathbb{R}),$$

then

$$\lim_{\delta \to 0+} \int_{|v|<\delta} |v'| \, dx = 0.$$

Proof We write

$$v_\delta = |v'| \chi_{\{|v|<\delta\}}, \qquad \delta > 0$$

and observe that

$$|v_\delta| \le |v'|, \qquad v_\delta \longrightarrow 0 \quad \text{a.e. in } \mathbb{R}.$$

Indeed, if $|\{v = 0\}| = 0$ we have $\chi_{\{|v|<\delta\}} \to 0$ otherwise $v' \to 0$ on $\{v = 0\}$. Therefore the claim follows from the Dominated Convergence Theorem. $\qquad\square$

Remark 6.4.1 Since the solutions of (6.61) are smooth, the previous lemma allows us to use the identity

$$\text{sign}\,(v)' = \delta_{\{v=0\}} v' \tag{6.67}$$

in our computations, where $\delta_{\{v=0\}}$ is the Dirac delta concentrated on the set $\{v = 0\}$. In particular, if $v \in C^2(\mathbb{R}) \cap L^\infty(\mathbb{R}) \cap W^{2,1}(\mathbb{R})$,

$$\int_{\mathbb{R}} f(v)'' \text{sign}\,(v') \, dx = 0, \qquad \int_{\mathbb{R}} v'' \text{sign}\,(v) \, dx \le 0, \tag{6.68}$$

that follow integrating by parts and using (6.67).

Let us give a rigorous proof of them. We have

$$\lim_{\alpha \to 0} \int_{\mathbb{R}} f(v)'' \eta'_\alpha(v') dx = \int_{\mathbb{R}} f(v)'' \text{sign}\,(v') \, dx,$$

$$\lim_{\alpha \to 0} \int_{\mathbb{R}} v'' \eta'_\alpha(v) dx = \int_{\mathbb{R}} v'' \text{sign}\,(v) \, dx, \tag{6.69}$$

where

$$\eta_\alpha(\xi) = \sqrt{\xi^2 + \alpha^2}, \qquad \alpha \in \mathbb{R}.$$

For every $\alpha \neq 0$

$$\eta_\alpha \in C^2(\mathbb{R}), \qquad \eta_\alpha'(\xi) = \frac{\xi}{\sqrt{\xi^2 + \alpha^2}}, \qquad \eta_\alpha''(\xi) = \frac{\alpha^2}{(\xi^2 + \alpha^2)^{3/2}} \geq 0.$$

We have

$$\left| \int_\mathbb{R} f(v)'' \eta_\alpha'(v') dx \right| = \left| \int_\mathbb{R} f'(v) v' \eta_\alpha''(v') v'' dx \right|$$

$$\leq L \int_\mathbb{R} \left| v' \eta_\alpha''(v') v'' \right| dx = L \int_\mathbb{R} \left| \frac{\alpha^2 v' v''}{((v')^2 + \alpha^2)^{3/2}} \right| dx$$

$$= L \int_{\{|v'| < \sqrt{\alpha}\}} \left| \frac{\alpha^2 v' v''}{((v')^2 + \alpha^2)^{3/2}} \right| dx$$

$$+ L \int_{\{|v'| \geq \sqrt{\alpha}\}} \left| \frac{\alpha^2 v' v''}{((v')^2 + \alpha^2)^{3/2}} \right| dx$$

$$\leq \frac{3}{8} L \underbrace{\int_{\{|v'| < \sqrt{\alpha}\}} \left| v'' \right| dx}_{\to 0 \text{ by Lemma 6.4.1}}$$

$$+ L \underbrace{\frac{\alpha}{(1+\alpha)^{3/2}}}_{\to 0} \int_{\{|v'| \geq \sqrt{\alpha}\}} \left| v'' \right| dx \longrightarrow 0,$$

$$\int_\mathbb{R} v'' \eta_\alpha'(v) dx = - \int_\mathbb{R} \eta_\alpha''(v)(v')^2 dx \leq 0,$$

where $L = \sup\limits_{|\xi| \leq \|v\|_{L^\infty(\mathbb{R})}} |f'(\xi)|$. Therefore, (6.68) follows from (6.69).

Let us continue with some apriori estimates on u_ε independent on ε.

Lemma 6.4.2 (L^∞ Estimate)   *We have that*

$$\|u_\varepsilon\|_{L^\infty((0,\infty) \times \mathbb{R})} \leq \|u_0\|_{L^\infty(\mathbb{R})}, \qquad \varepsilon > 0.$$

Proof Due to (6.63) the maps with constant values $\|u_0\|_{L^\infty(\mathbb{R})}$ and $-\|u_0\|_{L^\infty(\mathbb{R})}$ provide a super and a sub solution to (6.61), respectively. Therefore, the claim follows from the comparison principle for parabolic equations.   □

Lemma 6.4.3 (L^1 **Estimate**)   *The function*

$$t \geq 0 \longmapsto \|u_\varepsilon(t, \cdot)\|_{L^1(\mathbb{R})}$$

is nonincreasing. In particular,

$$\|u_\varepsilon(t, \cdot)\|_{L^1(\mathbb{R})} \leq \|u_0\|_{L^1(\mathbb{R})}, \qquad \varepsilon > 0, \quad t \geq 0.$$

Proof Due to the regularity of u_ε, we have

$$\frac{d}{dt} \int_{\mathbb{R}} |u_\varepsilon| dx = \int_{\mathbb{R}} \text{sign}(u_\varepsilon) \, \partial_t u_\varepsilon dx$$

$$= \varepsilon \int_{\mathbb{R}} \text{sign}(u_\varepsilon) \, \partial_{xx}^2 u_\varepsilon dx - \int_{\mathbb{R}} \text{sign}(u_\varepsilon) \, f'(u_\varepsilon) \partial_x u_\varepsilon dx$$

$$= \underbrace{-\varepsilon \int_{\mathbb{R}} \delta_{\{u_\varepsilon = 0\}} (\partial_x u_\varepsilon)^2 \, dx}_{\leq 0}$$

$$\underbrace{- \int_{\mathbb{R}} \partial_x \left(\int_0^{u_\varepsilon(t,x)} \text{sign}(s) \, f'(s) ds \right) dx}_{=0} \leq 0,$$

where $\delta_{\{u_\varepsilon = 0\}}$ is the Dirac's delta concentrated on the set $\{u_\varepsilon = 0\}$. Finally, an integration on $(0, t)$ gives (see (6.63))

$$\|u_\varepsilon(t, \cdot)\|_{L^1(\mathbb{R})} \leq \|u_{0,\varepsilon}\|_{L^1(\mathbb{R})} \leq \|u_0\|_{L^1(\mathbb{R})}.$$

$\square$

Lemma 6.4.4 (BV **Estimate in** x)   *The function*

$$t \geq 0 \longmapsto \|\partial_x u_\varepsilon(t, \cdot)\|_{L^1(\mathbb{R})}$$

is nonincreasing. In particular,

$$\|\partial_x u_\varepsilon(t, \cdot)\|_{L^1(\mathbb{R})} \leq TV(u_0), \qquad \varepsilon > 0, \quad t \geq 0.$$

Proof Due to the regularity of u_ε, we have

$$\partial_{tx}^2 u_\varepsilon + \partial_x \left(f'(u_\varepsilon) \partial_x u_\varepsilon \right) = \varepsilon \partial_{xxx}^3 u_\varepsilon$$

and then

$$\frac{d}{dt} \int_{\mathbb{R}} |\partial_x u_\varepsilon| dx = \int_{\mathbb{R}} \text{sign} \, (\partial_x u_\varepsilon) \, \partial_{tx}^2 u_\varepsilon dx$$

$$= \varepsilon \int_{\mathbb{R}} \text{sign} \, (\partial_x u_\varepsilon) \, \partial_{xxx}^3 u_\varepsilon dx - \int_{\mathbb{R}} \text{sign} \, (\partial_x u_\varepsilon) \, \partial_x \left(f'(u_\varepsilon) \partial_x u_\varepsilon \right) dx$$

$$= \underbrace{-\varepsilon \int_{\mathbb{R}} \delta_{\{\partial_x u_\varepsilon = 0\}} \left(\partial_{xx}^2 u_\varepsilon \right)^2 dx}_{\leq 0} + \underbrace{\int_{\mathbb{R}} \delta_{\{\partial_x u_\varepsilon = 0\}} \partial_{xx}^2 u_\varepsilon f'(u_\varepsilon) \partial_x u_\varepsilon dx}_{=0} \leq 0,$$

where $\delta_{\{\partial_x u_\varepsilon = 0\}}$ is the Dirac's delta concentrated on the set $\{\partial_x u_\varepsilon = 0\}$. Finally, an integration on $(0, t)$ gives (see (6.63))

$$\|\partial_x u_\varepsilon(t, \cdot)\|_{L^1(\mathbb{R})} \leq \|\partial_x u_{0,\varepsilon}\|_{L^1(\mathbb{R})} \leq TV(u_0).$$

□

Lemma 6.4.5 (*BV* **Estimate in** t) *The function*

$$t \geq 0 \longmapsto \|\partial_t u_\varepsilon(t, \cdot)\|_{L^1(\mathbb{R})}$$

is nonincreasing. In particular,

$$\|\partial_t u_\varepsilon(t, \cdot)\|_{L^1(\mathbb{R})} \leq TV(u_0)L + C, \qquad \varepsilon > 0, \quad t \geq 0,$$

where C is the constant that appears in (6.63) and

$$L = \|f'\|_{L^\infty(-\|u_0\|_{L^\infty(\mathbb{R})}, \|u_0\|_{L^\infty(\mathbb{R})})} \, .$$

Proof Due to the regularity of u_ε, we have

$$\partial_{tt}^2 u_\varepsilon + \partial_x \left(f'(u_\varepsilon) \partial_t u_\varepsilon \right) = \varepsilon \partial_{txx}^3 u_\varepsilon$$

and then

$$\frac{d}{dt} \int_{\mathbb{R}} |\partial_t u_\varepsilon| dx$$

$$= \int_{\mathbb{R}} \text{sign} \, (\partial_t u_\varepsilon) \, \partial_{tt}^2 u_\varepsilon dx$$

$$= \varepsilon \int_{\mathbb{R}} \text{sign} \, (\partial_t u_\varepsilon) \, \partial_{txx}^3 u_\varepsilon dx - \int_{\mathbb{R}} \text{sign} \, (\partial_t u_\varepsilon) \, \partial_x \left(f'(u_\varepsilon) \partial_t u_\varepsilon \right) dx$$

$$= \underbrace{-\varepsilon \int_{\mathbb{R}} \delta_{\{\partial_t u_\varepsilon = 0\}} \left(\partial_{tx}^2 u_\varepsilon \right)^2 dx}_{\leq 0} + \underbrace{\int_{\mathbb{R}} \delta_{\{\partial_t u_\varepsilon = 0\}} \partial_{tx}^2 u_\varepsilon f'(u_\varepsilon) \partial_t u_\varepsilon dx}_{=0} \leq 0,$$

where $\delta_{\{\partial_t u_\varepsilon=0\}}$ is the Dirac's delta concentrated on the set $\{\partial_t u_\varepsilon = 0\}$. Finally, an integration on $(0, t)$, (6.61), (6.63), and Lemma 6.4.2 give

$$\|\partial_t u_\varepsilon(t, \cdot)\|_{L^1(\mathbb{R})} \leq \|\partial_t u_\varepsilon(0, \cdot)\|_{L^1(\mathbb{R})}$$

$$= \left\| \varepsilon\partial^2_{xx} u_{0,\varepsilon} - f'(u_{0,\varepsilon})\partial_x u_{0,\varepsilon} \right\|_{L^1(\mathbb{R})}$$

$$\leq \varepsilon \left\| \partial^2_{xx} u_{0,\varepsilon} \right\|_{L^1(\mathbb{R})} + \left\| f'(u_{0,\varepsilon}) \right\|_{L^\infty(\mathbb{R})} \left\| \partial_x u_{0,\varepsilon} \right\|_{L^1(\mathbb{R})}$$

$$\leq C + TV(u_0)L.$$

$\square$

Proof (of (6.64)) Let $\{u_{\varepsilon_k}\}_{k\in\mathbb{N}}$ be a subsequence of $\{u_\varepsilon\}_{\varepsilon>0}$. Since $\{u_{\varepsilon_k}\}_{k\in\mathbb{N}}$ is bounded in $L^\infty((0, \infty) \times \mathbb{R}) \cap BV((0, T) \times \mathbb{R})$, $T > 0$, (see Lemmas 6.4.3, 6.4.4, and 6.4.5), there exists a function $u \in L^\infty((0, \infty) \times \mathbb{R}) \cap BV((0, T) \times \mathbb{R})$, $T > 0$, and a subsequence $\{u_{\varepsilon_{k_h}}\}_{h\in\mathbb{N}}$ such that

$$u_{\varepsilon_{k_h}} \longrightarrow u \qquad \text{in } L^p_{loc}((0, \infty) \times \mathbb{R}) \text{ and a.e. in } (0, \infty) \times \mathbb{R}.$$

We claim that u is the unique entropy solution of (6.62). Let $\eta \in C^2(\mathbb{R})$ be a convex entropy with flux q defined by $q' = \eta' f'$. Multiplying (6.61) by $\eta'(u_{\varepsilon_{k_h}})$ we get

$$\partial_t \eta(u_{\varepsilon_{k_h}}) + \partial_x q(u_{\varepsilon_{k_h}}) = \varepsilon_{k_h}\partial^2_{xx} u_{\varepsilon_{k_h}} \eta'(u_{\varepsilon_{k_h}})$$

$$= \varepsilon_{k_h}\partial^2_{xx} \eta(u_{\varepsilon_{k_h}}) \underbrace{- \varepsilon_{k_h}\eta''(u_{\varepsilon_{k_h}})(\partial_x u_{\varepsilon_{k_h}})^2}_{\leq 0}$$

$$\leq \varepsilon_{k_h}\partial^2_{xx} \eta(u_{\varepsilon_{k_h}}).$$

For every nonnegative test function $\varphi \in C^\infty(\mathbb{R}^2)$ with compact support we have that

$$\int_0^\infty \int_{\mathbb{R}} \left(\eta(u_{\varepsilon_{k_h}})\partial_t\varphi + q(u_{\varepsilon_{k_h}})\partial_x\varphi \right) dt\,dx + \int_{\mathbb{R}} \eta(u_{0,\varepsilon_{k_h}}(x))\varphi(0, x)dx$$

$$\geq -\varepsilon_{k_h} \int_0^\infty \int_{\mathbb{R}} \eta(u_{\varepsilon_{k_h}})\partial^2_{xx}\varphi\,dt\,dx.$$

As $h \to \infty$, the Dominated Convergence Theorem gives

$$\int_0^\infty \int_{\mathbb{R}} (\eta(u)\partial_t\varphi + q(u)\partial_x\varphi)\, dt\,dx + \int_{\mathbb{R}} \eta(u_0(x))\varphi(0, x)dx \geq 0,$$

proving that u is the unique entropy solution of (6.62).

Finally, thanks to (6.66), (6.64) is proved.

$\square$

6.4.2   Error Estimate

In this section we complete the proof of Theorem 6.4.1 showing (6.65).

Let t, $\varepsilon > 0$. We "double the variables", using (τ, x) for (6.62) and (s, y) for (6.61). We have

$$
\partial_t |u(\tau, x) - u_\varepsilon(s, y)|
$$
$$
+ \partial_x[\text{sign}\,(u(\tau, x) - u_\varepsilon(s, y))\,(f(u(\tau, x)) - f(u_\varepsilon(s, y)))] \leq 0, \tag{6.70}
$$

and

$$
\partial_s |u(\tau, x) - u_\varepsilon(s, y)|
$$
$$
+ \partial_y[\text{sign}\,(u(\tau, x) - u_\varepsilon(s, y))\,(f(u(\tau, x)) - f(u_\varepsilon(s, y)))] \tag{6.71}
$$
$$
\leq \varepsilon\partial_{yy}^2 |u(\tau, x) - u_\varepsilon(s, y)|,
$$

in the sense of distributions. Let $w \in C^\infty(\mathbb{R})$ be a nonnegative function with compact support such that

$$
\|w\|_{L^1(\mathbb{R})} = 1.
$$

We define

$$
w_\alpha(\xi) = \frac{1}{\alpha} w\left(\frac{\xi}{\alpha}\right), \qquad \xi \in \mathbb{R}, \ \alpha > 0.
$$

By testing (6.70) with the function

$$
(\tau, x) \longmapsto w_\beta(\tau - s)w_\alpha(x - y), \qquad \alpha, \beta > 0,
$$

we get

$$
\int_0^t \int_\mathbb{R} \Big[|u(\tau, x) - u_\varepsilon(s, y)|w_\beta'(\tau - s)w_\alpha(x - y)
$$
$$
+ \text{sign}\,(u(\tau, x) - u_\varepsilon(s, y))\,(f(u(\tau, x)) - f(u_\varepsilon(s, y)))\times
$$
$$
\times\, w_\beta(\tau - s)w_\alpha'(x - y)\Big]d\tau dx
$$
$$
- \int_\mathbb{R} |u(t, x) - u_\varepsilon(s, y)|w_\beta(t - s)w_\alpha(x - y)dx
$$
$$
+ \int_\mathbb{R} |u_0(x) - u_\varepsilon(s, y)|w_\beta(-s)w_\alpha(x - y)dx \geq 0,
$$

that is

$$\int_0^t \int_{\mathbb{R}} \int_{\mathbb{R}} |u(t, x) - u_\varepsilon(s, y)| w_\beta(t - s) w_\alpha(x - y) ds dx dy$$

$$\leq \int_0^t \int_{\mathbb{R}} \int_{\mathbb{R}} |u_0(x) - u_\varepsilon(s, y)| w_\beta(-s) w_\alpha(x - y) ds dx dy$$

$$+ \int_0^t \int_0^t \int_{\mathbb{R}} \int_{\mathbb{R}} \Big[|u(\tau, x) - u_\varepsilon(s, y)| w_\beta'(\tau - s) w_\alpha(x - y) \tag{6.72}$$

$$+ \operatorname{sign}(u(\tau, x) - u_\varepsilon(s, y)) (f(u(\tau, x)) - f(u_\varepsilon(s, y))) \times$$

$$\times w_\beta(\tau - s) w_\alpha'(x - y) \Big] ds d\tau dx dy.$$

By testing (6.71) with the function

$$(s, y) \longmapsto w_\beta(\tau - s) w_\alpha(x - y), \qquad \alpha, \beta > 0,$$

we get

$$- \int_0^t \int_{\mathbb{R}} \Big[|u(\tau, x) - u_\varepsilon(s, y)| w_\beta'(\tau - s) w_\alpha(x - y)$$

$$+ \operatorname{sign}(u(\tau, x) - u_\varepsilon(s, y)) (f(u(\tau, x)) - f(u_\varepsilon(s, y))) \times$$

$$\times w_\beta(\tau - s) w_\alpha'(x - y) \Big] ds dy$$

$$- \int_{\mathbb{R}} |u(\tau, x) - u_\varepsilon(t, y)| w_\beta(\tau - t) w_\alpha(x - y) dy$$

$$+ \int_{\mathbb{R}} |u(\tau, x) - u_{0,\varepsilon}(y)| w_\beta(\tau) w_\alpha(x - y) dy$$

$$\geq - \varepsilon \int_0^t \int_{\mathbb{R}} |u(\tau, x) - u_\varepsilon(s, y)| w_\beta(\tau - s) w_\alpha''(x - y) ds dy,$$

that is

$$\int_0^t \int_{\mathbb{R}} \int_{\mathbb{R}} |u(\tau, x) - u_\varepsilon(t, y)| w_\beta(\tau - t) w_\alpha(x - y) d\tau dx dy$$

$$\leq \int_0^t \int_{\mathbb{R}} \int_{\mathbb{R}} |u(\tau, x) - u_{0,\varepsilon}(y)| w_\beta(\tau) w_\alpha(x - y) d\tau dx dy$$

$$- \int_0^t \int_0^t \int_{\mathbb{R}} \int_{\mathbb{R}} \Big[|u(\tau, x) - u_\varepsilon(s, y)| w_\beta'(\tau - s) w_\alpha(x - y)$$

$$+ \operatorname{sign}(u(\tau, x) - u_\varepsilon(s, y)) (f(u(\tau, x)) - f(u_\varepsilon(s, y))) \times$$

$$\times w_\beta(\tau - s) w_\alpha'(x - y) \Big] ds d\tau dx dy$$

$$+ \varepsilon \int_0^t \int_0^t \int_{\mathbb{R}} \int_{\mathbb{R}} |u(\tau, x) - u_\varepsilon(s, y)| w_\beta(\tau - s) w_\alpha''(x - y) ds d\tau dx dy.$$

$$\tag{6.73}$$

We add (6.72) and (6.73)

$$\int_0^t \int_{\mathbb{R}} \int_{\mathbb{R}} |u(t,x) - u_\varepsilon(s,y)| w_\beta(t-s) w_\alpha(x-y) \, ds \, dx \, dy$$

$$+ \int_0^t \int_{\mathbb{R}} \int_{\mathbb{R}} |u(\tau,x) - u_\varepsilon(t,y)| w_\beta(\tau - t) w_\alpha(x-y) \, d\tau \, dx \, dy$$

$$\leq \int_0^t \int_{\mathbb{R}} \int_{\mathbb{R}} |u_0(x) - u_\varepsilon(s,y)| w_\beta(-s) w_\alpha(x-y) \, ds \, dx \, dy$$

$$+ \int_0^t \int_{\mathbb{R}} \int_{\mathbb{R}} |u(\tau,x) - u_{0,\varepsilon}(y)| w_\beta(\tau) w_\alpha(x-y) \, d\tau \, dx \, dy$$

$$+ \varepsilon \int_0^t \int_0^t \int_{\mathbb{R}} \int_{\mathbb{R}} |u(\tau,x) - u_\varepsilon(s,y)| w_\beta(\tau - s) w_\alpha''(x-y) \, ds \, d\tau \, dx \, dy$$

and send $\beta \to 0$

$$\underbrace{\int_{\mathbb{R}} \int_{\mathbb{R}} |u(t,x) - u_\varepsilon(t,y)| w_\alpha(x-y) \, dx \, dy}_{I_1}$$

$$\leq \underbrace{\int_{\mathbb{R}} \int_{\mathbb{R}} |u_0(x) - u_{0,\varepsilon}(y)| w_\alpha(x-y) \, dx \, dy}_{I_2} \qquad (6.74)$$

$$+ \underbrace{\frac{\varepsilon}{2} \int_0^t \int_{\mathbb{R}} \int_{\mathbb{R}} |u(s,x) - u_\varepsilon(s,y)| w_\alpha''(x-y) \, ds \, dx \, dy}_{I_3}.$$

We estimate I_1 and I_2 in the following way (see (6.63) and Lemma 6.4.4)

$$I_1 \geq \int_{\mathbb{R}} \int_{\mathbb{R}} \left(|u(t,x) - u_\varepsilon(t,x)| - |u_\varepsilon(t,x) - u_\varepsilon(t,y)| \right) w_\alpha(x-y) \, dx \, dy$$

$$= \int_{\mathbb{R}} |u(t,x) - u_\varepsilon(t,x)| \, dx - \int_{\mathbb{R}} \int_{\mathbb{R}} |u_\varepsilon(t, y+\xi) - u_\varepsilon(t,y)| w_\alpha(\xi) \, d\xi \, dy$$

$$\geq \|u(t,\cdot) - u_\varepsilon(t,\cdot)\|_{L^1(\mathbb{R})} - \int_{\mathbb{R}} \left| \int_0^\xi \int_{\mathbb{R}} |\partial_x u_\varepsilon(t, y+\sigma)| \, dy \, d\sigma \right| w_\alpha(\xi) \, d\xi$$

$$= \|u(t,\cdot) - u_\varepsilon(t,\cdot)\|_{L^1(\mathbb{R})} - \|\partial_x u_\varepsilon(t,\cdot)\|_{L^1(\mathbb{R})} \int_{\mathbb{R}} |\xi| w_\alpha(\xi) \, d\xi$$

$$\geq \|u(t,\cdot) - u_\varepsilon(t,\cdot)\|_{L^1(\mathbb{R})} - \alpha TV(u_0) \int_{\mathbb{R}} |\xi| w(\xi) \, d\xi,$$

$$I_2 \leq \int_{\mathbb{R}} \int_{\mathbb{R}} \left(|u_0(x) - u_{0,\varepsilon}(x)| + |u_{0,\varepsilon}(x) - u_{0,\varepsilon}(y)| \right) w_\alpha(x-y) \, dx \, dy$$

$$= \int_{\mathbb{R}} |u_0(x) - u_{0,\varepsilon}(x)| dy + \int_{\mathbb{R}} \int_{\mathbb{R}} |u_{0,\varepsilon}(y + \xi) - u_{0,\varepsilon}(y)| w_\alpha(\xi) d\xi dy$$

$$\leq \|u_0 - u_{0,\varepsilon}\|_{L^1(\mathbb{R})} + \int_{\mathbb{R}} \left| \int_0^\xi \int_{\mathbb{R}} |\partial_x u_{0,\varepsilon}(y + \sigma)| dy d\sigma \right| w_\alpha(\xi) d\xi$$

$$= \|u_0 - u_{0,\varepsilon}\|_{L^1(\mathbb{R})} + \|\partial_x u_{0,\varepsilon}\|_{L^1(\mathbb{R})} \int_{\mathbb{R}} |\xi| w_\alpha(\xi) d\xi$$

$$\leq \|u_0 - u_{0,\varepsilon}\|_{L^1(\mathbb{R})} + \alpha T V(u_0) \int_{\mathbb{R}} |\xi| w(\xi) d\xi.$$

We have to estimate I_3. Thanks to (6.64) we know

$$I_3 = \lim_{\mu \to 0} I_{3,\mu},$$

where

$$I_{3,\mu} = \frac{\varepsilon}{2} \int_0^t \int_{\mathbb{R}} \int_{\mathbb{R}} |u_\mu(s, x) - u_\varepsilon(s, y)| w_\alpha''(x - y) ds dx dy, \qquad \mu > 0.$$

Since (see Lemma 6.4.4)

$$I_{3,\mu} \leq \frac{\varepsilon}{2} \int_0^t \int_{\mathbb{R}} \int_{\mathbb{R}} \left(|\partial_x u_\mu(s, x)| + |\partial_y u_\varepsilon(s, y)| \right) |w_\alpha'(x - y)| ds dx dy$$

$$= \frac{\varepsilon}{2} \int_0^t \int_{\mathbb{R}} \int_{\mathbb{R}} \left(|\partial_x u_\mu(s, y + \xi)| + |\partial_y u_\varepsilon(s, y)| \right) |w_\alpha'(\xi)| ds d\xi dy$$

$$= \frac{\varepsilon}{2} \int_0^t \int_{\mathbb{R}} \left(\|\partial_x u_\mu(s, \cdot)\|_{L^1(\mathbb{R})} + \|\partial_y u_\varepsilon(s, \cdot)\|_{L^1(\mathbb{R})} \right) |w_\alpha'(\xi)| ds d\xi$$

$$\leq \varepsilon t T V(u_0) \|w_\alpha'\|_{L^1(\mathbb{R})} = \frac{\varepsilon t}{\alpha} T V(u_0) \|w'\|_{L^1(\mathbb{R})},$$

we have

$$I_3 \leq \frac{\varepsilon t}{\alpha} T V(u_0) \|w'\|_{L^1(\mathbb{R})}.$$

Using the estimates on I_1, I_2, and I_3 in (6.74) we have

$$\|u(t, \cdot) - u_\varepsilon(t, \cdot)\|_{L^1(\mathbb{R})} \leq \|u_0 - u_{0,\varepsilon}\|_{L^1(\mathbb{R})}$$

$$+ \left(\alpha + \frac{\varepsilon t}{\alpha} \right) T V(u_0) \left(2 \int_{\mathbb{R}} |\xi| w(\xi) d\xi + \|w'\|_{L^1(\mathbb{R})} \right).$$

Since the minimum of the map

$$\alpha \longmapsto \alpha + \frac{\varepsilon t}{\alpha}$$

is attained in $\sqrt{\varepsilon t}$, (6.65) is proved.

Appendix: BV Functions

In this section we collect some elementary facts about functions with bounded variations since their relevance in the study of conservation laws.

Definition 6.4.1 Let $I \subset \mathbb{R}$ be an interval and let $u : I \to \mathbb{R}$. The total variation of f over I is defined by

$$TV(u) = \sup \sum_{k=0}^{q} |u(t_{k+1}) - u(t_k)| \qquad (6.75)$$

where the supremum is taken over all finite sequences $t_0 < \ldots < t_q$ so that $t_i \in I$, for every i. The function u is said to be of *bounded variation* on I, in symbol $u \in BV(I)$, if $TV(u) < \infty$. It is easy to verify that the sum of two functions of bounded variations is also of bounded variation. Before proving the converse, let us introduce the notation $V_u(a; x)$ to denote the total variation of the function u on the interval (a, x). Observe that if u is of bounded variation on $[a, b]$ and $x \in [a, b]$, then

$$|u(x) - u(a)| \leq V_u(a; x) \leq V_u(a; b) = TV(u).$$

Theorem 6.4.2 *If u is a function of bounded variation on $[a, b]$, then u can be written as*

$$u = u_1 - u_2$$

where u_1 and u_2 are nondecreasing functions.

Proof Let $x_1 < x_2 \leq b$ and let $a = t_0 < t_1 < \ldots < t_k = x_1$. Then

$$V_u(x_2) \geq |u(x_2) - u(x_1)| + \sum_{i=1}^{k} |u(t_i) - u(t_{i-1})|.$$

Since by definition

$$V_u(x_1) = \sup \sum_{i=1}^{k} |u(t_i) - u(t_{i-1})|$$

over all the sequences $a = t_0 < t_1 < \ldots t_k = x_1$, we get

$$V_u(x_2) \geq |u(x_2) - u(x_1)| + V_u(x_1).$$

Therefore

$$V_u(x_2) - u(x_2) \geq V_u(x_1) - u(x_1), \quad V_u(x_2) + u(x_2) \geq V_u(x_1) + u(x_1).$$

Hence $V_u - u$ and $V_u + u$ are nondecreasing functions. The claim follows by taking

$$u_1 = \frac{1}{2}(V_u + u), \quad u_2 = \frac{1}{2}(V_u - u).$$

$\square$

Theorem 6.4.3 *Let u be a function of bounded variation on* $[a, b]$. *Then u is Borel measurable and has at most a countable number of discontinuities. Moreover, the following statements hold true*

(i) u' *exists a.e. on* $[a, b]$;
(ii) u' *is Lebesgue measurable;*
(iii) for a.e. $x \in [a, b]$

$$|u'(x)| = V'_u(x);$$

(iv)

$$\int_a^b |u'(x)|\, dx \le V_u(b);$$

(v) if u is nondecreasing on $[a, b]$, *then*

$$\int_a^b u'(x)\, dx \le u(b) - u(a).$$

The following theorem due to Helly is a fundamental result in the theory of bounded variation functions.

Theorem 6.4.4 *Let* $u_n : [a, b] \to \mathbb{R}$ *be a sequence of functions satisfying the condition*

$$\sup_n TV(u_n) < \infty. \tag{6.76}$$

Then there exists a subsequence, still denoted by u_n *and a function u of bounded variation such that* $u_n(x) \to u(x)$ *as* $n \to \infty$ *for every* $x \in [a, b]$ *and*

$$TV(u) \le \liminf_n TV(u_n). \tag{6.77}$$

Acknowledgments The authors are members of the Gruppo Nazionale per l'Analisi Matematica, la Probabilità e le loro Applicazioni (GNAMPA) of the Istituto Nazionale di Alta Matematica (INdAM).

Lecture notes of the course "Conservation Laws in Continuum Mechanics" held by GMC in Cetraro (CS) on July 1–5, 2019 during the CIME-EMS Summer School in applied mathematics "Applied Mathematical Problems in Geophysics".

References

1. A. Aw, M. Rascle, Resurrection of "second order" models of traffic flow. SIAM J. Appl. Math. **60**(3), 916–938 (2000)
2. C. Bardos, A.Y. le Roux, J.-C. Nédélec, First order quasilinear equations with boundary conditions. Comm. Partial Differ. Equ. **4**(9), 1017–1034 (1979)
3. S. Benzoni-Gavage, D. Serre, *Multidimensional Hyperbolic Partial Differential Equations*. Oxford Mathematical Monographs (The Clarendon Press, Oxford University Press, Oxford, 2007). First-order systems and applications
4. S. Bianchini, A. Bressan, Vanishing viscosity solutions of nonlinear hyperbolic systems. Ann. Math. (2) **161**(1), 223–342 (2005)
5. A. Bressan, *Hyperbolic Systems of Conservation Laws*, vol. 20. Oxford Lecture Series in Mathematics and its Applications (Oxford University Press, Oxford, 2000). The one-dimensional Cauchy problem
6. C.M. Dafermos, *Hyperbolic Conservation Laws in Continuum Physics*, vol. 325. *Grundlehren der Mathematischen Wissenschaften [Fundamental Principles of Mathematical Sciences]*, 2nd edn. (Springer, Berlin, 2005)
7. A. Friedman, *Partial Differential Equations* (Robert E. Krieger Publishing Co., Huntington, 1976). Original edition
8. M. Garavello, B. Piccoli, *Traffic Flow on Networks*, vol. 1. AIMS Series on Applied Mathematics (American Institute of Mathematical Sciences (AIMS), Springfield, 2006). Conservation laws models
9. M. Garavello, K. Han, B. Piccoli, *Models for Vehicular Traffic on Networks*, vol. 9. AIMS Series on Applied Mathematics (American Institute of Mathematical Sciences (AIMS), Springfield, 2016)
10. S.K. Godunov, E.I. Romenskii, *Elements of Continuum Mechanics and Conservation Laws* (Kluwer Academic/Plenum Publishers, New York, 2003). Translated from the 1998 Russian edition by Tamara Rozhkovskaya
11. H. Holden, N.H. Risebro, *Front Tracking for Hyperbolic Conservation Laws*, vol. 152. Applied Mathematical Sciences, 2nd edn. (Springer, Heidelberg, 2015)
12. S.N. Kružkov, First order quasilinear equations with several independent variables. Mat. Sb. (N.S.) **81**(123), 228–255 (1970)
13. N.N. Kuznecov, The accuracy of certain approximate methods for the computation of weak solutions of a first order quasilinear equation. Ž. Vyčisl. Mat. i Mat. Fiz. **16**(6), 1489–1502, 1627 (1976)
14. O.A. Ladyženskaja, V.A. Solonnikov, N.N. Ural'ceva, *Linear and Quasilinear Equations of Parabolic Type*. Translated from the Russian by S. Smith. Translations of Mathematical Monographs, vol. 23 (American Mathematical Society, Providence, 1968)
15. M.J. Lighthill, G.B. Whitham, On kinematic waves. I. Flood movement in long rivers. Proc. Roy. Soc. London. Ser. A. **229**, 281–316 (1955)
16. P. Prasad, *Nonlinear Hyperbolic Waves in Multi-Dimensions*, vol. 121. Chapman & Hall/CRC Monographs and Surveys in Pure and Applied Mathematics (Chapman & Hall/CRC, Boca Raton, 2001)
17. P.I. Richards, Shock waves on the highway. Oper. Res. **4**, 42–51 (1956)
18. D. Serre, *Systems of Conservation Laws. 1* (Cambridge University Press, Cambridge, 1999) Hyperbolicity, entropies, shock waves, Translated from the 1996 French original by I. N. Sneddon
19. D. Serre, *Systems of Conservation Laws. 2* (Cambridge University Press, Cambridge, 2000) Geometric structures, oscillations, and initial-boundary value problems, Translated from the 1996 French original by I. N. Sneddon
20. V.D. Sharma, *Quasilinear Hyperbolic Systems, Compressible Flows, and Waves*, vol. 142. Chapman & Hall/CRC Monographs and Surveys in Pure and Applied Mathematics (CRC Press, Boca Raton, 2010)
21. L. Tartar, *From Hyperbolic Systems to Kinetic Theory*, vol. 6. Lecture Notes of the Unione Matematica Italiana (Springer, Berlin; UMI, Bologna, 2008). A personalized quest

LECTURE NOTES IN MATHEMATICS  Springer

Editors in Chief: J.-M. Morel, B. Teissier;

Editorial Policy

1. Lecture Notes aim to report new developments in all areas of mathematics and their applications – quickly, informally and at a high level. Mathematical texts analysing new developments in modelling and numerical simulation are welcome.

   Manuscripts should be reasonably self-contained and rounded off. Thus they may, and often will, present not only results of the author but also related work by other people. They may be based on specialised lecture courses. Furthermore, the manuscripts should provide sufficient motivation, examples and applications. This clearly distinguishes Lecture Notes from journal articles or technical reports which normally are very concise. Articles intended for a journal but too long to be accepted by most journals, usually do not have this "lecture notes" character. For similar reasons it is unusual for doctoral theses to be accepted for the Lecture Notes series, though habilitation theses may be appropriate.

2. Besides monographs, multi-author manuscripts resulting from SUMMER SCHOOLS or similar INTENSIVE COURSES are welcome, provided their objective was held to present an active mathematical topic to an audience at the beginning or intermediate graduate level (a list of participants should be provided).

   The resulting manuscript should not be just a collection of course notes, but should require advance planning and coordination among the main lecturers. The subject matter should dictate the structure of the book. This structure should be motivated and explained in a scientific introduction, and the notation, references, index and formulation of results should be, if possible, unified by the editors. Each contribution should have an abstract and an introduction referring to the other contributions. In other words, more preparatory work must go into a multi-authored volume than simply assembling a disparate collection of papers, communicated at the event.

3. Manuscripts should be submitted either online at www.editorialmanager.com/lnm to Springer's mathematics editorial in Heidelberg, or electronically to one of the series editors. Authors should be aware that incomplete or insufficiently close-to-final manuscripts almost always result in longer refereeing times and nevertheless unclear referees' recommendations, making further refereeing of a final draft necessary. The strict minimum amount of material that will be considered should include a detailed outline describing the planned contents of each chapter, a bibliography and several sample chapters. Parallel submission of a manuscript to another publisher while under consideration for LNM is not acceptable and can lead to rejection.

4. In general, **monographs** will be sent out to at least 2 external referees for evaluation.

   A final decision to publish can be made only on the basis of the complete manuscript, however a refereeing process leading to a preliminary decision can be based on a pre-final or incomplete manuscript.

   Volume Editors of **multi-author works** are expected to arrange for the refereeing, to the usual scientific standards, of the individual contributions. If the resulting reports can be forwarded to the LNM Editorial Board, this is very helpful. If no reports are forwarded or if other questions remain unclear in respect of homogeneity etc, the series editors may wish to consult external referees for an overall evaluation of the volume.

5. Manuscripts should in general be submitted in English. Final manuscripts should contain at least 100 pages of mathematical text and should always include

   – a table of contents;
   – an informative introduction, with adequate motivation and perhaps some historical remarks: it should be accessible to a reader not intimately familiar with the topic treated;
   – a subject index: as a rule this is genuinely helpful for the reader.
   – For evaluation purposes, manuscripts should be submitted as pdf files.

6. Careful preparation of the manuscripts will help keep production time short besides ensuring satisfactory appearance of the finished book in print and online. After acceptance of the manuscript authors will be asked to prepare the final LaTeX source files (see LaTeX templates online: https://www.springer.com/gb/authors-editors/book-authors-editors/manuscriptpreparation/5636) plus the corresponding pdf- or zipped ps-file. The LaTeX source files are essential for producing the full-text online version of the book, see http://link.springer.com/bookseries/304 for the existing online volumes of LNM). The technical production of a Lecture Notes volume takes approximately 12 weeks. Additional instructions, if necessary, are available on request from lnm@springer.com.

7. Authors receive a total of 30 free copies of their volume and free access to their book on SpringerLink, but no royalties. They are entitled to a discount of 33.3 % on the price of Springer books purchased for their personal use, if ordering directly from Springer.

8. Commitment to publish is made by a *Publishing Agreement*; contributing authors of multiauthor books are requested to sign a *Consent to Publish form*. Springer-Verlag registers the copyright for each volume. Authors are free to reuse material contained in their LNM volumes in later publications: a brief written (or e-mail) request for formal permission is sufficient.

Addresses:
Professor Jean-Michel Morel, CMLA, École Normale Supérieure de Cachan, France
E-mail: moreljeanmichel@gmail.com

Professor Bernard Teissier, Equipe Géométrie et Dynamique,
Institut de Mathématiques de Jussieu – Paris Rive Gauche, Paris, France
E-mail: bernard.teissier@imj-prg.fr

Springer: Ute McCrory, Mathematics, Heidelberg, Germany,
E-mail: lnm@springer.com

Printed in the United States
by Baker & Taylor Publisher Services